The Cambridge Photographic Atlas of the Planets

The Cambridge

Photographic Atlas of the Planets

GEOFFREY BRIGGS

Deputy Director, Earth and Planetary Exploration Program, NASA

FREDRIC TAYLOR

Acting Head, Department of Atmospheric Physics
University of Oxford

CAMBRIDGE UNIVERSITY PRESS

Cambridge

London New York New Rochelle

Melbourne Sydney

Published by the Press Syndicate of the University of Cambridge
The Pitt Building, Trumpington Street, Cambridge CB2 1RP
32 East 57th Street, New York, NY 10022, USA
296 Beaconsfield Parade, Middle Park, Melbourne 3206, Australia

First published 1982

Printed in Singapore

Library of Congress catalogue card number: 81–38529

British Library cataloguing in publication data
The Cambridge photographic atlas of the planets.
Briggs, Geoffrey
1. Planets
I. Title II. Taylor, Fredric
523.4   QB601
ISBN 0 521 23976 1

Contents

Preface

Although the exploration of the Solar System is still in its earliest phase we have been able, with the successful Voyager flybys of Jupiter and Saturn, to celebrate recently the acquisition of close-up views of all of the planets that were known to the ancient world: over the past twenty years we have come to know about a score of new worlds, including the larger natural satellites of the outer planets and our own Moon. These worlds have proven to be fascinating, even bizarre, beyond any of our expectations. During the two decades in question men have walked on the Moon (and have provided us with an arresting perspective of our home planet from the depths of space), and robot spacecraft have landed on Mars and Venus, while orbital and flyby probes have encountered each of the planets out to Saturn. In most cases cameras have been a basic part of the scientific payloads, providing planetary scientists with a versatile investigatory technique and allowing, also, the world at large to participate in an inspiring adventure comparable to the voyages of discovery to the New World in the fifteenth and sixteenth centuries.

Many tens of thousands of planetary images have been returned to Earth from space for analysis by geologists and atmospheric scientists. Because most of the other solid planetary surfaces, unlike the Earth, still record the major evolutionary episodes through which the body has passed since its original formation, it has proved possible to read directly the history of these bodies and to attach a relative timescale to the stages of evolution. A fundamental tool in such analyses is the map, prepared from the photographic coverage with great skill by airbrush artists working under strict geodetic control. In this atlas we have selected about two hundred of the best planetary images and these are presented together with appropriate maps of the planets and their moons – maps that are generally global in scale (many larger scale maps have also been prepared by the planetary mapmakers at the U.S. Geological Survey in Flagstaff, Arizona, for the use of professional photogeologists but it is not possible to include them here).

Accompanying the images and maps in this atlas (and not all the images are made from video data – some use radar and others use infrared data) are captions and a text which attempts to summarize what we have learned about the planets to date. We have not attempted to provide a detailed historical perspective of the science nor of the missions, nor have we provided references to the scientific literature in this atlas which is intended for the general reader. We therefore use this opportunity to acknowledge the work of all our colleagues that is described broadly herein. Certain of our colleagues provided general help in reviewing parts of the text of this book and especial thanks are due to William Quaide, Michael Carr, Laurel Wilkening, and Joseph Boyce. Nearly all of the pictures contained in this atlas were acquired by spacecraft launched by the U.S. National Aeronautics and Space Administration. Considerable help has been provided to us in locating and acquiring pictures from the NASA files and we wish to thank the following individuals for their help on this and other matters in the assembling of this volume: Dave Diner, Ken Carroll, Andy Collins, Tom Duxbury, Nancy Evans, Les Gaver, Elizabeth Newhouse, Larry Soderblom, Dick Underwood, Kay Voglewede and Jeff Warner. The infrared picture of Jupiter is by Richard Terrile. The pictures of Io on page 198 were computer processed at the U.S. Geological Survey in Flagstaff and were provided by Larry Soderblom. For the Venera pictures we acknowledge the Novosti Press Agency. Except where specifically mentioned, the maps are the product of the Astrogeology Branch of the U.S. Geological Survey. We wish to thank Mr. Ray Batson for his help in acquiring these maps. The map of the Earth on pages 90–1 was prepared by Miss Marie Tharp, and copies may be obtained from her at 1 Washington Avenue, South Nyack, N.Y. 10960, U.S.A.

Introduction

Celestial objects, whether viewed unaided in the night sky, or seen through telescopes and in the pictures returned by spacecraft, are of compelling beauty. To the scientist they pose problems of a fundamental nature which, little by little, are yielding to his enquiry. The study of the Solar System with its nine planets, numerous satellites, asteroids and comets together with its central star, the Sun, is an area where stellar astronomy, astrophysics and planetary astronomy meet. In recent years, owing to the space programmes of the larger nations and the stimulus of these programmes to astronomers and theorists throughout the world, a considerable amount of progress has been made in Solar System studies, and some of the nearer planets have indeed come to seem like our close neighbours.

The nine planets are notable for their individual, widely varying characteristics, but they also show similarities that allow them to be classified into broad groups. The planets of the inner Solar System – Mercury, Venus, Earth and Moon, and Mars – are all solid objects that may be described as *rocky* in general constitution. The asteroids, in orbits which generally lie between those of Mars and Jupiter, also fit this description. Beyond Mars are the planets of the outer Solar System – Jupiter, Saturn, Uranus and Neptune. The first two of these appear to have compositions similar to the Sun and, although the gaseous phase is not maintained at great depth, are termed *gaseous* planets. The higher densities of Uranus and Neptune clearly indicate that they are not twins of Jupiter and Saturn, but they too fit this general class. Furthest out of the planets, Pluto revolves around the Sun only four times in a millennium. Discovered in 1920, Pluto largely eludes the probing of even the most powerful Earth-based telescopes and, thereby, maintains its mystery.

Spending most of their lives far beyond even the orbit of Pluto are the comets – tiny bodies of ice and dust suspended in a permanent deep freeze. Occasionally the trajectory of a comet will be disturbed by the gravity field of a nearby star so that the comet is nudged into an orbit that brings it into the inner Solar System. Here, the heat of the Sun causes the icy material of the cometary 'nucleus' to sublime so that this tiny object produces a tenuous atmosphere of gases and dust, together with extended tails of dust and of ionized gases. It is in this moment of glory that comets can become the most spectacular objects in the sky.

Until quite recently the planets, asteroids and comets received essentially all of the attention of planetary astronomers: the moons of planets other than our own were largely overlooked because so little could be learned about them with available techniques. This situation has changed greatly now as telescopic and spacecraft capabilities have increased. The four large Galilean moons of Jupiter and Saturn's largest moon, Titan, are now studied by astronomers as intensively as the planets. In size these moons are comparable to the Moon and Mercury and they exhibit an extraordinary diversity so that the intense interest that they have provoked among planetary scientists is well founded.

When NASA's Space Telescope becomes operational in the mid-1980s, and when planetary spacecraft reach beyond Saturn, we can expect that, not only will Uranus, Neptune and Pluto begin to reveal their secrets, but also that the moons of these objects will become characterized and understood.

Origin of the Solar System

The measured abundances of the isotopes produced by the radioactive decay of certain chemical elements can be used to determine the age of naturally occurring materials. Based on the radioactive dating of meteorites and of lunar samples, there is ample evidence that the Solar System is about 4.6 billion years old. The Universe as a whole is thought to be between 10 and 20 billion years old and to have been steadily expanding ever since its origin at the time of the so-called 'Big Bang'. Subsequent to that cataclysmic event, many generations of stars have been formed and have evolved and died, with the result that the Universe has undergone some important changes. From the parochial view of an Earthling, a vital result of the Universe's evolution has been the creation of chemical elements more massive than hydrogen and helium (the simplest and lightest of all elements and the only ones present from the beginning). Without these heavier elements a planet like the Earth, whose most common elements include iron, magnesium, silicon and oxygen, could not have formed. The heavier elements are believed to be formed by the fusion of lighter elements in the interiors of relatively massive stars, and also by more complex nuclear processes in supernovae (the explosive death throes of such massive stars).

As the Universe has evolved, increasing quantities of the heavier elements have been created and dispersed among the galaxies by supernova explosions. Most of the mass of a galaxy still remains as the lightest elements – hydrogen and helium – which are contained within stars and as gas within the interstellar medium. Only a very tiny part of the interstellar medium is in the form of heavier elements which are believed to be found mostly within tiny dust grains, 10^{-8}–10^{-6} metres in size. Spectroscopic measurements indicate that these particles are chiefly made of carbon, in its graphite form, of water ice, and of silicates of magnesium and iron. Most of the dust and gas appears to be in the arms of spiral galaxies, like the one in which our own Sun resides and where most new stars are formed.

Stars are formed, generally in clusters, when a local region of gas and dust becomes sufficiently dense – by random chance or, perhaps, as a result of a specific event such as a nearby supernova – that the gravitational forces of attraction between the matter win out over the internal pressure of the gas. The material begins to collapse. As the collapse continues the material apparently divides into fragments, each of which ultimately becomes a star (or often a binary star system) – perhaps with an associated planetary system which is a minor byproduct of star formation. As a result of turbulence within the material each fragment will have some net angular momentum and hence will be subject to rotation. As the dimensions of each fragment become smaller the rate of rotation increases so that the collapsing fragment spins out to form a flattened disc which is termed a *solar nebula*. At the centre of the nebula, where temperatures and pressures are highest, is the *protosun*.

It is supposed that our Sun and all the other objects within the Solar System formed from such a solar nebula. The details of the process of formation represent a problem that lies at the heart of planetary science and provides an important part of the motivation behind our exploration of the planets. This exploration has already revealed important systematic compositional differences between the planets – differences that provide fundamental constraints on theories of Solar System formation and evolution.

Much of what we presently know about the Solar System has been derived, not from data returned by spacecraft, but from delicate analyses of meteorites– samples of extraterrestrial material that impact the Earth and can be investigated using all the sophisticated techniques available in the modern laboratory. Meteorites are derived from many different parent bodies and range in compositional type from almost pure nickel–iron to mixtures of apparently primitive minerals. The latter type of meteorite is known as a *carbonaceous chondrite* and typically consists of tiny pieces of igneous minerals embedded in a matrix of low-temperature silicate minerals. These meteorites are thought to be examples of quite primitive solar nebula material, and studies of their composition and mineralogy have been invaluable in providing information about the earliest history of the Solar System. In the past few years analyses of minute inclusions within carbonaceous chondrites have provided evidence of the anomalous presence of two relatively short-lived radioactive elements (isotopes of aluminium and of palladium) that are now long extinct. These two nuclides have half-lives (the time it takes for 50% of the atoms to undergo radioactive decay) of only

about a million years, the implication of the discovery being that the isotopes must have been created just before the formation of the Solar System. It has been suggested that a supernova may have occurred at this time, creating these elements (among others) and that this newly created material did not have sufficient time to mix with the bulk of the material that was to form the solar nebula – if such mixing *had* occurred then the entire nebula would have had a uniform average composition and no anomalies would be detectable. If a supernova did in fact take place at the time of Solar System formation then it seems quite likely that this event was the trigger for the collapse of the dust and gas that comprised the solar nebula.

Formation of the planets

How did the solar nebula evolve to form the Sun and planets? At the present time this is still the big open question, and there is a range of theories from which to choose. None provides a completely satisfying answer since each can only explain part of the observational data provided by meteorite analysis, planetary astronomy and spacecraft exploration. The theoretical models will be described only briefly since the details are evolving rapidly and soon become obsolete.

First, a major constraint upon models of Solar System origin will be described, namely the compositional differences that exist between the objects in the Solar System. The gross difference between the inner, rocky planets and the outer, gaseous planets has already been mentioned. In addition, there are significant, systematic differences among the inner and the outer planets which are known from determinations of the densities of these bodies (derived from mass and size measurements). In comparing the densities of the planets, and in trying to relate the densities to bulk composition, proper allowance must be made for the fact that the larger a planet is the greater the compression it exerts on the material in its interior. When this allowance is made a marked density gradient is observed for the inner planets from Mercury (mean 'uncompressed' density, $5.4\,\mathrm{g/cm^3}$) to Mars ($3.35\,\mathrm{g/cm^3}$). The Earth and Venus have similar values at about $4.2\,\mathrm{g/cm^3}$. It has been concluded that this density gradient must imply a systematic variation in the amount of iron (a cosmically abundant element that is much heavier than other common elements) that has been incorporated in the inner planets. Any model of Solar System origin must be able to explain this observation.

Among the four giant outer planets the density trend reverses, and it has been concluded that the amount of the lightest elements, hydrogen and helium, decreases as one moves outwards beyond Jupiter. Thus, where Jupiter is thought to be composed of over 80% hydrogen and helium, Saturn has only about 70%, Uranus 15%, and Neptune 10%. In the latter two cases the composition is thought to comprise mainly water, ammonia and methane. Models of Solar System origin must also be compatible with these observations.

The concept of a rotating, disc-like nebula is one that is common to all models that are being seriously considered at this time. Such a concept leads naturally to a systematic variation in temperature and pressure throughout the nebula, with the highest values near the centre where the Sun formed. There is, however, considerable uncertainty about what those temperatures and pressures were at the different locations where each of the nine planets formed. Some estimates suggest that a temperature difference of as much as 1000°C existed between the orbit of proto-Mercury and the orbit of proto-Mars. The dust component of the solar nebula may have been vaporized in some of its inner regions. The recondensation of such a vapour, or simply the equilibration of unvaporized nebula material, would follow a well-defined mineral sequence as a function of temperature, given a certain original composition (a composition exemplified by that of carbonaceous chondrite meteorites). Certainly, material equilibrating and subsequently accreting into larger bodies at high temperatures will contain proportionally more iron than material equilibrating at low temperature. We may, therefore, understand the gradient in the amount of iron contained in the inner planets as being the result of a radial temperature gradient in the solar nebula, but this general understanding is subject to numerous problems when the details are considered. It is in the details that the various Solar System models depart from one another.

Working out the details of how the solar nebula condensed and how a plethora of small condensations (mostly silicate materials in the inner Solar System, a mixture of silicates and ices in the outer Solar System) grew into a few large planetary-sized objects is proving to be a formidable problem, involving both physical and chemical processes. The physical aspects include the processes by which tiny grains of condensed material collide and accumulate into increasingly larger bodies: such processes are complex and by no means well understood. Current models suggest that the process of accumulation into objects of asteroidal dimensions proceeded rapidly – a time-scale measured in thousands of years. These 'planetismals' (up to tens of kilometres in size) are thought to have been the building blocks from which the inner planets were made.

Although there must have been innumerable planetismals formed, and many are still located in the asteroid belt where the gravitational effects of Jupiter prevent the accretion of a solid planet, it is clear that the final collisional processes leading to planet formation must have greatly favoured accretion by large objects such that, ultimately, the planetismals were swept up to form only four planets in the inner Solar System. The advantage of a large gravity field in sweeping up material is clear – collisions lead to the production of debris moving at high velocity which can be recaptured by a big object but which will be lost to space by a small one.

A very large amount of heat must have been generated by the planetismal impacts that led to the accretion of the planets, and the bigger the planet the more the heating because the larger gravity field involved would have increased the velocity of impact. How much of this heat was radiated away from the growing body and how much was retained to increase the temperature of the body is an unresolved question. If the process of accretion took place sufficiently quickly then enough heat may have been retained to have led to substantial melting of the inner planets. In the case of the Moon there is evidence in the returned lunar samples that the lunar surface was originally molten to a depth of several hundred kilometres, presumably because of this early gravitational heating. Larger planetary bodies like the Earth might have been even more extensively melted.

In the outer Solar System, where grains of both silicate materials and of water, ammonia and methane ices are thought to have been available, the growth of the two bodies that ultimately became Jupiter and Saturn may have proceeded sufficiently rapidly that massive objects were formed that were able to gravitationally attract large quantities of the nebula gases (hydrogen and helium). This would account for their observed bulk compositions and would imply that these planets have cores of silicaceous materials. Further out, in the vicinity of Neptune and Uranus, the protoplanets perhaps grew less rapidly and were less efficient in capturing the nebular gases.

It is thought that the solar nebula was several times more massive than the present Sun, where all but a tiny fraction of the mass of the Solar System resides. Therefore, it appears that a large part of the solar nebula must have been lost. Based on astronomical observations of the behaviour of young stars elsewhere in the Galaxy, it is thought that the early Sun was once much more massive, but that it passed through a stage of vigorous activity during which an intense solar 'wind' blew outwards through the Solar System taking with it much of the mass of the youthful Sun and also sweeping away

Introduction

gaseous material that had not yet condensed or been accumulated onto a growing planetary object. This event, known as the *T Tauri* phase of the Sun, probably took place during the late stages of planetary accretion, following which the Solar System probably existed in the general form that we see today. Since that time each of the planets has followed its own individual evolution to its present state, a state that we are, at last, able to investigate using the eyes and other senses of planetary spacecraft.

Evolution of the planets

As might be expected, the evolutionary path along which a planetary body starts depends upon the category – rocky or gaseous – to which it belongs. In the case of Jupiter and Saturn the original accumulation of dust and gas must have occupied a significantly larger volume than at present because originally these objects were much hotter and their internal pressures greater. Theoretical models of these planets treat them as small stars and follow their history using techniques originally developed to study stellar evolution. There is one vital difference, of course, that simplifies the models – internal temperatures sufficient for hydrogen fusion are never reached and the body remains a planet. The early stages of contraction of the gaseous planets are comparatively rapid such that a body ten times greater in diameter than at present reaches a size roughly double present size in between a hundred thousand and a million years. From that point the rate of contraction slows as a result of the build up of internal pressure and the following decrease in diameter to that observed today occupies several billion years. The luminosity of the planet increases greatly during the contraction phase as gravitational energy is released and some is radiated from the surface. Not all of this energy escapes, however, and very high internal temperatures are generated. During the final phase of contraction heat loss is greater than that produced by continued contraction so that the planet slowly cools – the phase in which both Jupiter and Saturn appear to be today. Telescopic and spacecraft measurements show that both planets are emitting significantly more heat than they are receiving from the Sun – the excess is believed to represent the continuing loss of heat 'buried' in the earliest contraction phase. In the case of Saturn the gravitational migration of helium towards the centre of the planet is also thought to be a key factor. Theoretical models suggest that internal temperatures are too high for solids to be found inside Jupiter or Saturn though the hydrogen component near the centre is thought to assume the conduction properties of a metal under the extreme

temperature and pressure conditions encountered there. Once their initial hot contraction phase had been completed these planets have probably undergone rather little subsequent evolution and may be among the least changed of all the bodies in the Solar System.

The evolution of the inner, rocky planets has been entirely different, though again the earliest stages were probably the most dramatic. The violence of the early life of these planets is written clearly on the surfaces of Mercury, the Moon and Mars and there is no reason to suppose that Venus or the Earth was spared the steady rain of infalling bodies. In the case of the Moon, where unlike the Earth crustal material from the early period has been preserved, the minerals contained in the returned lunar samples suggest that the surface was molten to a considerable depth. The heat of planetary accretion due to infalling asteroidal material and heat resulting from the decay of short-lived radioactive elements may both have been involved in creating this spectacular episode. We cannot look for clues to the nature of the Earth's earliest history on the greatly modified surface of our planet, but observations of other planetary surfaces and the return of samples from other bodies may provide sufficient evidence to allow us to infer the conditions on Earth at the dawn of its existence.

Significant atmospheric *outgassing* (release of gases from the interior to form an atmosphere) may have taken place very early in the history of the inner planets since the presumed early episode of heating could have led to the dissociation of minerals containing chemically bound volatile materials such as water and carbon dioxide. An early atmosphere could have suffered the effects of the T Tauri solar wind if the relative timing of events was right. Evidence of massive sweeping of a planetary atmosphere may be assembled from detailed studies of the compositions of the atmospheres of Mars and Venus. Volatiles not outgassed during the period of planetary accretion may have been released into the atmosphere later after radiogenic heat had built up inside the planet. The history of atmospheric outgassing is closely tied to the thermal evolution of the planet in question and, therefore, since the latter is a basic area of investigation, the study of a planet's inventory of gases, and of condensed and chemically bound volatiles, is given high priority in the planning of planetary spacecraft missions.

Atmospheric outgassing to form a gravitationally bound gaseous shell about the body is perhaps one of the earliest *differentiations* that occurs in a planet's life. Similar separations between materials of differing densities can also occur within the body to produce an internal zonation: an extensively differentiated inner (or *terrestrial*) planet might have a *core* of very dense, mostly metallic material at its centre, a *mantle* of metallic silicate material surrounding the core, and a *crust* of the lightest minerals (also generally silicates) at the surface. The outermost zone, the atmosphere, may in time be lost or modified in a variety of ways as will be discussed later.

In some models of planet formation the interior of a planet is directly separated into zones of differing densities by virtue of the sequential accretion of materials of different mineral composition and of different density. In other models the accreted planet is initially of fairly uniform composition, and internal differentiation comes later as a result of the thermal evolution of the body. To achieve the differentiation described the planet needs, together with its gravity field which separates materials of differing densities, a source of heat to first melt the rocks. Two sources of heat are probably the most important: gravitational (including the heat of accretion) and radiogenic (which can operate over all time scales including billions of years). Each of the planets would have been subject to such heating and it is supposed on the basis of both observation and theoretical analysis that all of the terrestrial planets have experienced some degree of differentiation.

The first major internal differentiation may have been the core formation. The heaviest material that is incorporated into a planet in large quantities is iron. When internal temperatures had, as a result of the intense early heating and the accumulation of sufficient radiogenic heat, reached its melting point, then quantities of iron would fall to the centre of the planet. Iron is readily combined chemically with oxygen and with sulphur so that, given sufficient quantities of these elements in the original mix, there might be no free iron in a planet. In that event, instead of a core of metallic iron, a core of iron sulphide might be produced. The separation of a core, if it occurs, is itself an important source of gravitational heat as the heaviest material completes its fall to the centre of the planet. Since this is an internal event that traps all the heat involved it is a very efficient form of gravitational heating.

Access to the deep interiors of solid planets is denied to us except through seismic measurements. Therefore, most of what we have learned thus far about how the other solid planets have evolved has been through the study of their surfaces. Kilometre-resolution global-scale imaging data, which is now available for all the inner planets except Venus, have been of particular importance in providing a general and graphic means for elucidating the major events that have shaped the history of a planet. In later chapters the individual characteristics of the planets, and their histories as we presently understand them, will be described. Below, a general description of the nature of the surfaces of solid planets will be provided as background for the later discussions.

The surfaces of solid planets

We have found that the evolutionary paths of the various planetary bodies explored to date have diverged greatly. These paths have produced landscapes quite unlike those with which we are familiar on Earth. However, if allowance is made for the lack of oceans and for the lack of vegetation, certain common basic ingredients in the nature of the landscapes can be discerned. There are four such ingredients: impact cratering; tectonism (crustal fracturing) and volcanism; and gradation (the smoothing of the landscape due to weathering and erosion). The first three processes *create* different types of surface terrain while the fourth tends to *erase* them. Each will be discussed briefly, making some broad comparisons between the planets.

Impact cratering

As far out from the Sun as Saturn there is clear evidence that all of the planets were subject to an intensive bombardment of meteoritic material around the time of their formation. This bombardment has led to the formation of impact craters. On a number of planets and moons such craters dominate the appearance of the surface. Where this is so, the body in question has reached the stage of senility in its evolution very rapidly so that the original surface has not been eradicated by later global-scale events as on the Earth. Evidence for ancient (and modern) impact events on the Earth is, however, quite plentiful although ancient terrestrial craters have all been so greatly modified that they are rarely obvious.

Impact craters come in all sizes, ranging up to hundreds of kilometres in diameter. Typically the impacts took place at relative velocities of many kilometres per second. At such velocities the size of the crater is many times that of the impacting body. For example, a kilometre-sized crater of a hundred metres depth could be created by the impact of a body only a few tens of metres in diameter.

The cratering process in rocky materials (as opposed to the icy materials from which the crusts of most of the outer planet satellites appear to be formed) is quite well understood: most of the destruction is caused by the passage of high-pressure shock waves produced at the moment of impact. Melting and vaporization do occur but are not major contributors to the excavation. Typically, material flows out of the forming crater for several seconds and creates what is called an *ejecta blanket* around the crater, stretching out for several crater diameters. Some of the ejected material, travelling at a suitable angle and with high velocity, may be spread over very large distances.

Apart from the size and velocity of the incoming missile, the chief factors controlling the sizes and shapes of craters are the strength of the surface material (in the extreme case of zero strength as in an ocean surface no crater will be formed) and the strength of the planet's gravity: measurable differences between crater characteristics on the Moon and on Mercury are attributable to the greater strength of Mercury's gravity which prevents the ejection of material over distances as large as on the Moon.

Craters smaller than a few kilometres in size tend to have a characteristic bowl shape. Larger craters are more flattened and exhibit terrace walls as well as hills at their centres. These central peaks are absent from the largest craters which measure hundreds of kilometres across. They are termed *basins* and are characterized by multiple, concentric rings of mountains created when the impacted material subsequently relaxed.

The most severe surface modification due to meteoritic bombardment occurred soon after the final stage of planet formation. By this time the surface, if originally melted, had cooled to a point where impacts could be recorded. At this time impacts were sufficiently numerous to *saturate* the surfaces in craters such that a new crater, in forming, partially or completely obliterated earlier craters. In time the impact rate tapered off and today is very low – but is not zero. Arizona's famous Barringer Crater (also called Meteor Crater) is thought to have been formed about 20 000 years ago. Probably there are many thousands of objects similar in size (tens of metres) to the one that created Barringer Crater (about a kilometre across) in orbits that intersect the Earth's. Impacts are rare but spectacular. In the case of a collision with an asteroid a few kilometres in diameter the results would be catastrophic on a global scale. It has been suggested that the extinction of the dinosaurs and other species 65 million years ago, at the boundary between the Cretaceous and Tertiary eras, may have been the result of secondary effects, global in scale, that followed an asteroidal impact.

Some large objects that impact the Earth may not be dense enough to survive the passage through the atmosphere to the surface and may not produce a crater. The so-called Tunguska event in Siberia in 1908 is notable for the absence of a crater, in spite of the widespread destruction of forest land that accompanied the event. The impacting object may have been a small comet.

In addition to their importance as evidence of conditions in the early Solar System, impact craters are studied for information about the relative timing of surface-forming events on a planet. The flux of

impacting bodies has continued since the planets were formed and cratering is not limited to the earliest-formed terrains. A large volcanic flood occurring, say, a billion years after a planet formed will cover all the previously formed craters but will then begin to accumulate a crater population, of lower areal density, of its own. Therefore, measurements of the areal density of impact craters can provide information about the relative chronology of events.

Tectonism

Tectonism is the name applied to processes that cause fracturing of the crust of a planet: such processes, clearly, are the result of colossal forces whose magnitude cannot be related to anything in our ordinary experience. Crustal fracturing on a planet-wide scale has been observed on many objects. It is present on Mars and Mercury, on the Galilean satellites Europa and Ganymede, and probably is to be found on Venus. Tectonic activity on a global scale is fundamental to the geology of the Earth.

Conceptually, the simplest source of a force that could create global fracturing is the uniform expansion or contraction of an entire planet due to the heating or cooling of its interior. To be most effective, such changes would need to occur relatively rapidly on the scale of geologic time. The formation of the core of a planet or the change in mineral phase of the planet's mantle material might cause relatively rapid expansion or contraction. Tidal forces acting throughout the interior of a planet as a result of the gravitational interaction of two or more bodies might also cause major crustal stresses.

Tectonism on Earth at the present time is evidently the result of forces different from those mentioned above. We know that the interior of the Earth is still very hot and it is thought that a very slow but inexorable convective overturning of the upper mantle is taking place that carries with it the overlying crust. As a result the Earth's crust has been broken apart and is now made up of a dozen irregular pieces, each called a *plate*. The continents are embedded in some of these plates and the geography of the world is therefore continually changing. A full appreciation of the reality of plate tectonics has only been achieved in the last few decades although the concept had been proposed much earlier – this new appreciation has entirely revolutionized our geological understanding of the Earth.

Until recently the Earth was the only planet known to have a multi-plate crust. The Voyager pictures of Jupiter's moon Ganymede have provided some evidence that this body also has a crust that has been broken into pieces. In the case of

Ganymede it is thought that the crust is composed of water ice which overlies a mantle of liquid water. It is supposed that convection in this mantle, resulting from heat produced in its presumed rocky core, has led to the fracturing.

One other planet remains a prime candidate to show some kind of global tectonism – our nearest neighbour, Venus. Being of similar size to the Earth, and having an apparently similar composition, Venus ought to have experienced a stage where plate tectonics could occur. To date the combined evidence of Earth-based radar measurements and those taken by the Pioneer Venus orbiter indicate that Venus has a substantially different surface morphology from that of the Earth. The state of the evolution of Venus remains one of the most important unresolved questions in the exploration of the planets.

Volcanism
Volcanism and tectonism are closely related phenomena that are hard to separate since they are both manifestations of a planet's internal heat engine. However, whereas tectonism has been discussed as a global process, volcanism will be treated as a more localized phenomenon, of which there are many varieties.

Perhaps the simplest style of volcanism is that observed on the Moon and Mercury, where huge plains of solidified lava are found. The lava – liquid rock created at depth by melting – was evidently extruded through elongated vents in the crust. The lava flowed freely across a broad area to create a plain rather than a mountain as would have been the case if the lava had been erupted through a localized vent. On the Moon these volcanic plains fill many of the basins. As such they are aptly given the Latin name for 'sea' – *mare* (plural *maria*). The precise location of the surface vent cannot be determined because it is buried by the flood. Chemical analyses of samples of lunar *mare* material reveal a *basaltic* composition indicative of melting at some depth where the available material consists of iron and magnesium silicates. Basaltic lavas tend to flow freely, and the extensive flooding can therefore be readily understood. (The steep conical form of terrestrial stratovolcanoes is the result of a central source of lava which has an *andesitic* composition – less iron and magnesium and more lightweight metals such as aluminium – and is more viscous. The molten rock is derived from continental crust rather than mantle material.)

On Mars volcanic flooding is also observed, but the most prominent, and spectacular, volcanoes there are huge, gently sloping domes termed *shield* volcanoes. Similar constructs are also found on Earth – the Hawaiian Islands, for example. The shield form is associated with basaltic lavas and on

Mars individual flows on the flanks of some of the shields can be traced for hundreds of kilometres. It seems fairly safe to suppose that the Martian shields are basaltic and the result of melting at great depth. The Martian shields can be several hundred kilometres across and, in four cases, they reach heights of well over 20 kilometres. A possible explanation for the great height may be that each volcano lies over a 'plume' (an ascending column of molten rock) which pumped lava to the surface from great depth over an extended period of time. Given a sufficiently thick and strong crust, the continued outpouring of lava could create a giant mountain whose height would be limited by the hydrostatic pressure that was forcing the lava upwards. This pressure would be determined by the depth of the lava source. This mechanism is probably responsible for the mountainous features on Venus, also.

In the case of the Earth, which is an extremely active planet geologically, all types of volcanism can be found including floods like those on the Moon and shields like those on Mars. Most familiar, however, are the graceful stratovolcanoes (e.g. Ranier, Vesuvius, Fujiyama): these are generally formed in chains at the edge of a continental plate as a consequence of the motion of the plates. The most common form of terrestrial volcanism is, paradoxically, the least familiar because it occurs in the middle of the oceans where molten mantle material is continuously extruded from the Earth's interior along enormous cracks – cracks which mark the boundaries between crustal plates. This accumulation of new basaltic crust is the result of the global-scale overturning of the upper mantle which causes the plates to move apart along some of their boundaries and to collide at other boundaries. The basalt is formed by partial melting of the mantle and rises to the surface as a result of convective motion.

One other, quite extraordinary, style of planetary volcanism – this time global in nature – must be mentioned in this brief discussion: that discovered on the Jovian moon Io. This satellite, the innermost of the four large Galilean moons, must surely exhibit the most active volcanism in the Solar System: hundreds of vents are observed across the globe and half a dozen or more eruptions appear to occur continuously. Internal heat due to radioactive decay cannot possibly account for such activity and, instead, it seems most likely that frictional heating, as a result of tides raised by the combined effects of Jupiter and the other Galilean satellites, is the cause.

Gradation
On planetary bodies without sensible atmospheres (the Moon, Mercury, and the Galilean satellites) the landscapes that are observed can be understood in

terms of the combined effects of cratering, tectonism and volcanism. On bodies with significant atmospheres (the Earth, Mars, Venus and Titan) the landscape is continually being modified as a result of erosion and deposition: the high areas are gradually lowered and the low areas filled in, in a process of landscape smoothing that is termed gradation.

Gradation involves three stages: first the weathering of material to soften it up, then the removal of material by some agent of transportation, and finally the deposition of the material when the transporting agent lacks sufficient energy to move it further. The force of gravity is fundamental to the transportation process: material is carried from high to low regions. The weathering of rocks may involve chemical or physical processes and will take place even on an airless body as a result of daily temperature changes, the blast of the solar wind, and the continuous impact of micrometeorites. However, on an airless planet transportation can only take place as a result of gravity: landslides occur but do not alter the general landscape to any great extent.

On a planet with an atmosphere, and especially on one with a condensible atmospheric constituent like water, transportation can be effected by wind action or fluvial action or glacial action. Because the Earth has such an abundance of liquid water, gradation through fluvial action is predominant in shaping the landscape. On Mars, orbital pictures show that fluvial erosion took place in the past. Such erosion may have been the result of catastrophic flooding rather than the drainage of rainfall and may not imply that Mars ever had an atmosphere akin to our own. Certainly today the atmosphere of Mars is so thin that liquid water cannot exist on the surface of Mars and fluvial erosion is no longer an active force.

The most important erosional process acting on Mars at the present time is the result of wind action. We can tell from pictures of Mars, taken from orbit and from the surface, that *aeolian* processes are still very important: global-scale dust storms occur regularly on Mars and serve to redistribute material about the planet. We can see that the higher latitudes have been mantled in debris at the expense of the more equatorial latitudes, and at both poles there are large expanses of smooth, uncratered (and therefore young) sedimentary terrain that have evidently been emplaced as a result of the airborne transportation of dust. In the north the polar sedimentary terrain is encircled by a vast sea of enormous dunes that clearly demonstrates the movement of material by the wind. Also at the Martian poles are permanent glaciers (water ice in the north and possibly carbon dioxide ice in the south) but erosional features related to glacial action have yet to be identified.

We lack the data required to understand the situation on Venus, a planet with an enormously thick atmosphere but only low-velocity surface winds. Little can be learned about gradational processes from the ground-based radar data or from the Pioneer Venus radar results, although they do show that the surface of Venus has substantial vertical relief. Two pictures of the surface of Venus have been returned by Soviet spacecraft and they reveal numerous sharp-edged rocks strewn about the surface. Gradational processes on Venus have clearly not produced a billiard ball surface.

The atmosphere of the planets

All of the planets except Mercury, and perhaps Pluto, have atmospheres. Most of the moons do not – they are generally too small to be able to retain an atmosphere throughout geologic time – but at least one does, namely, Titan, the largest of Saturn's moons. Like all their other characteristics, the atmospheres of the planets show wide variability as a result of the great range in planetary compositions, sizes and distances from the Sun at the time of formation.

The atmospheres of the planets have been studied extensively by instruments, because atmospheric behaviour is crucial in determining the surface environment, while its composition gives many clues to the evolution of the planet. Moreover, atmospheres are well suited for remote investigation: many common atmospheric gases (e.g. carbon dioxide and water vapour) emit and absorb visible, *ultraviolet* (wavelength shorter than visible) and *infrared* (wavelength longer than visible) radiation with a *spectrum* (variation with wavelength) that is characteristic not only of their composition but also of their temperature and abundance. On a global scale such measurements can be made using instruments placed at the focus of large telescopes, while from spacecraft measurements can be made with much better spatial resolution and made into maps which allow the weather and climate of the planet to be studied and compared to the Earth.

In this brief introduction to the subject, three aspects of planetary atmospheres will be discussed – evolution, thermal state and circulation.

Evolution
The compositions of planetary atmospheres vary widely. As has already been discussed briefly, the atmospheres of the giant outer planets appear to be largely unchanged since their formation, although the thermal state of each planet is certainly believed to have evolved significantly in the first few million years. Once that early thermal transient had passed, evolution has apparently been minimal and the composition has probably changed little. The

principal reason for this lies in the great size of these planets and in their remoteness from the Sun. The combination of low upper atmospheric temperatures and enormous gravitational fields prevents any significant loss of even the lightest gases. The case is quite different, however, for the inner planets which are warmer and have much weaker gravity and which are undergoing continual internal evolution with the consequent exhalation of gases.

Venus, Earth and Mars have atmospheric compositions that differ widely from the composition of the sun and from that of the primitive solar nebula. In these cases we must think in terms of the evolution of *secondary* atmospheres rather than the retention of *primary* atmospheres captured at the time of planet formation. Perhaps all the inner planets once had primary atmospheres that were *reducing* (i.e. hydrogen-rich and oxygen-poor) in character: there is little information available to help us decide. Such information as exists is provided by the measured abundances of rare gases like argon, krypton, neon and xenon which are both heavy and inert so that they are retained throughout geologic time with relatively little change. These measurements, made by Viking and Pioneer Venus, show very large differences between Mars, Earth and Venus but no uniform trend for all of the rare gases. Thus, rather than providing an immediate clarification of the nature of the formation and evolution of the planets, the data provide a fresh challenge.

It is supposed that little remains of any primary atmosphere for Venus, Earth and Mars and that even the earliest outgassed atmosphere was probably subject to sweeping during the Sun's T Tauri phase. However, it is likely that not all of the gases would have been lost. Furthermore, outgassing has probably continued throughout the entire history of the planet so that, in time, a substantial atmosphere might be accumulated and might subsequently evolve as a result of a variety of processes. In the case of Mercury, and also our own Moon, a weak gravity field has prevented the retention of either a primary or a secondary atmosphere. Also, their relatively small sizes (resulting in fairly rapid cooling of the interior) and their apparent deficiencies in volatile materials have probably resulted in very little outgassing on these bodies.

Both observation and theory suggest that the most abundant gases that would be released into the atmosphere as a result of outgassing would be water vapour, carbon dioxide and nitrogen. The relative proportions of these would be expected to be different for Venus, Earth and Mars because of their different original compositions. The absolute amounts of the gases that would be released from

the interiors would depend on the thermal history of each planet. Thus we would expect that these three bodies would inevitably have somewhat dissimilar atmospheres. The observed differences are, however, so great that they cannot be explained without considering the complex processes that are involved after the outgassing has taken place. These include the release of gases from the crystalization of molten rock and from radioactive decay processes, the breakup of molecules (for example as a result of photodissociation by sunlight), the loss of gases by chemical interaction with surface materials, and the escape to space of the lightest gases. In the case of Earth, biological processes have also been of crucial importance.

Distance from the Sun is vital, too: at Earth's distance water vapour has condensed on the surface in both liquid and solid form; Venus is extremely dry; Mars probably has large quantities of fresh water beneath its surface; on Titan methane appears to play a similar role to that of water on Earth, being available in gaseous, liquid, and solid phases.

Thermal state
The average vertical temperature structure in an atmosphere is determined mainly by the balance which exists at each level between incoming solar and outgoing thermal infrared (heat) radiation. Most of the Sun's radiant energy is at relatively short wavelengths near the visible part of the spectrum, overlapping into the ultraviolet on the short wave-length side and the *near* infrared on the long wave-length side. At these wavelengths, most atmospheric gases do not absorb very much. On the other hand, the outgoing planetary heat radiation is emitted mostly as photons of much lower energy which have longer wavelengths, in the *middle* and *far* infrared part of the spectrum. These are absorbed strongly by atmospheric gases such as carbon dioxide and water vapour, methane and ammonia, because the energies of the photons match the transition energies between the internal vibrational and rotational states of these molecules. The amount of infrared absorption over a given distance increases with increasing pressure as more molecules are available to absorb. The amount of *visible* absorption, however, stays small at all pressures of interest in the planetary atmospheres known to us. Thus, at increasingly greater depths in any atmosphere, heat radiation finds it more and more difficult to escape to cold space but sunlight continues to penetrate. This results in higher temperatures with increasing depth, since these increase the cooling rate to the point where equilibrium is achieved. In general, then, the higher the atmospheric pressure at any level (the surface, for example) the greater the temperature. Other factors which come into play are the solar intensity at the top of the atmosphere

(the distance of the planet from the Sun), the composition of the atmosphere (the ratio of *infrared-active* gases, like carbon dioxide and water vapour, to relatively inactive gases like nitrogen, oxygen and helium), and the role played by clouds. Clouds reflect, transmit, and absorb radiation of all wavelengths in a very complex manner. Also, the various levels in the atmosphere tend to exchange heat between themselves by advection as well as radiation. Finally, the role of heating from inside the planet must be considered. This is small compared to solar heating for the terrestrial planets, but large for the gas giants.

The thermal states of the three terrestrial planetary atmospheres make interesting comparisons. Mars has a thin atmosphere of carbon dioxide which interacts with incoming solar radiation hardly at all, and is only moderately opaque on average to outgoing heat energy. The net effect is that the surface temperature at which balance is obtained is only 10 °C or so higher than it would be if Mars had no atmosphere at all. At the other extreme, Venus has 10 000 times more carbon dioxide molecules in a vertical path than Mars and much more water vapour, too. With that much absorbing gas in the way, the surface cannot cool by radiation to space at all. Instead, less efficient convective processes predominate, taking heat to higher levels where it can radiate away. At the same time, heat is being supplied to the surface by sunlight diffusing through the clouds, relatively unimpeded by the transparent atmosphere. This state of affairs reaches a balance at the remarkably high surface temperature of nearly 460 °C.

The Earth is an intermediate case. It has less CO_2 than Mars but much more water vapour, and it also has a fairly thick stratospheric ozone layer, which neither Mars nor Venus has. Ozone is a good infrared absorber at some wavelengths, and is also a powerful absorber of (solar) ultraviolet rays. The latter effect makes the middle atmosphere of Earth very warm at pressures around 1 mb, around 0 °C compared to −100 °C on Venus and −90 °C on Mars. The mean temperature of the surface on Earth is about +20 °C.

The fractional cloud cover on Earth is intermediate between that of Venus (almost total) and that of Mars (almost zero) and so its variability is particularly important. The amount of cloud at any particular time is controlled by various atmospheric *feedback mechanisms*. A simple example of a feedback mechanism is provided by the case where sunlight impinges directly on to the ocean, evaporating water vapour until the atmosphere becomes saturated and clouds form. These shield the water surface from direct sunlight and cause its temperature to drop, tending to reduce evaporation and suppress further cloud formation. This is negative feedback. An example of positive feedback is the melting of ice or snow. Partial melting reduces the reflectivity of the surface by exposing underlying soil or rock; this allows a greater fraction of the incident sunlight to be absorbed which melts the snow more rapidly. The maintenance of the Earth's present climate (or that of the other planets) is a balance between many such mechanisms working with and against each other. In order to understand how stable the balance is, and how likely it is to alter either naturally or due to man's influence (which, through pollution, etc., is now considerable) requires a detailed unravelling of the physical and chemical processes at work. It is very useful when attempting to do this to intercompare observations of the terrestrial planets, since the relative importance of a given process is different on each.

Circulation

A planet's atmosphere is in continuous motion as a result of non-uniform radiative heating from the Sun. The non-uniformity has two principal components: equator–pole and day–night. The magnitudes of the dynamical forcing resulting from this imbalance depend on the rotation rate of the planet and the density of the atmosphere at the level being considered. For example, on Venus the atmosphere near the surface has nearly the same temperature everywhere because it is so dense, and hence such a good heat reservoir, that it would take many years for an increase in solar heating to raise the temperature by even a small amount. Before this happens, the diurnal cycle has averaged out the heating in the zonal direction, while extremely sluggish meridional motions have mixed the equator and polar air enough to suppress temperature gradients of any great size. In marked contrast, the upper atmosphere of Venus exhibits huge day–night temperature differences and high winds. Here the rarified atmosphere heats very rapidly when illuminated by the Sun, and the associated pressure differences result in motions which tend – but at these levels never successfully – to eliminate the temperature difference. To the simple picture described above must be added the effects of *Coriolis forces,* which modify the motions on any spinning planet, and axial tilt, which introduces seasonal effects. Surface features introduce localized heating rate gradients and also interrupt the flow at the lowest levels, sometimes triggering wave motions. Waves of various kinds occur over a wide range of time and spatial scales in most atmospheres and can strongly modify the general circulation. Finally, albedo effects (including vegetation and ice on the Earth) and transport of heat in the oceans exert a major effect on the heat distribution and hence the circulation in our own atmosphere.

Venus has perhaps the simplest general

circulation of all of the planets which have been explored in any detail. This is because of its very slow rotation rate and near-absence of axial tilt or orbital eccentricity (hence no seasonal variations). The motions seem to be dominated by convective overturning between equator and pole. On Earth such large-scale convective cells are not dynamically stable and the circulation is characterized by mid-latitude jet streams and smaller cells (global scale eddies). Mars in its winter hemisphere tends to resemble the Earth but in its summer hemisphere is more like Venus.

The circulation of the outer, gas giant planets is further complicated by the presence of internal heat sources which rival or exceed the solar input. Very strong convection would be expected in this situation and it is this, combined with the rapid rotation of these planets, which produces the characteristic banded appearance.

Although our understanding of the dynamics of planetary atmospheres has advanced by large strides in the last two decades, we remain well aware of how much is left to be learned and understood. It has become something of a cliché to describe the planets as natural laboratories for all kinds of atmospheric and solid-body geoscience investigations. Nevertheless, the brief history of planetary exploration to date amply supports this point of view.

Mercury

As a result of being the innermost of the planets and, therefore, always in a difficult position to observe near the Sun, Mercury was little known until March 1974 when the Mariner 10 spacecraft made the first of its three flybys of that body. The general nature of Mercury as a relatively unevolved lunar-like body is now recognized. The basic physical characteristics, as determined mainly by optical and radar telescope measurements, are summarized in the following table:

Mass 3.28×10^{26} g	Orbital period 88 days
Radius (equatorial) 2440 km	Obliquity $\sim 0°$
Mean density 5.5 g/cm³	Orbital eccentricity 0.206
Equatorial gravity 368 cm/s²	Mean distance from
Rotational period 59 days	the Sun 57.9×10^6 km

Radar telescope data were particularly important in pinning down the rotation period (day) which is, as closely as can be determined, two-thirds of the orbital period (year). Optical telescopic observers had been misled for many years into concluding that the orbital and rotation periods were identical; that is, that Mercury's rotation had become 'locked' into 1:1 synchronism with its orbital period – like our Moon's situation with respect to the Earth. An interesting result of the observed 3:2 ratio between the two periods is that it takes 176 terrestrial days (two orbital periods) for the Mercurian 'day' to be completed. Moreover, Mercury's orbit is more eccentric than any of the other planets except Pluto, so that the intensity of sunlight on the surface varies by about a factor of two between closest approach to the Sun (perihelion) and furthest (aphelion). If Mercury's rotational and orbital periods had been the same, then the same region on the planet would always have faced the Sun at perihelion – a 'hot pole'. The observed 3:2 resonance leads to the alternation of the 'hot seat' between two regions on opposite sides of the planet – longitudes 0° and 180° – which, therefore, provide Mercury with two hot poles.

A fundamental characteristic of any planet is its density, which provides the strongest constraint on a basic property that cannot be measured directly – the bulk composition of the planet. The density is determined from measurements of the mass of the planet and its diameter. In interpreting the calculated density, allowance must be made for the size of the body because of the effects of what may be called 'self-compression' in its interior; i.e. the increase in density caused by the pressure of the overlying material. Mercury, although much smaller than the Earth, has a similar density – about 5.5 g/cm³. Common silicate rocks have a density of about 3 g/cm³ and self-compression of such rocky material cannot account for a density of 5.5 g/cm³ for either Mercury or the Earth. A major additional component of some much heavier material is required. Based on our knowledge of the cosmic abundance of the elements (derived from observations of the Sun's composition and from analysis of certain meteorites) it is evident that the most likely candidate heavy element is iron. Indeed, we have learned from seismic measurements that the Earth has a core with characteristics attributable to a mainly iron composition. Given the small size of Mercury (which provides less self-compression) it is evident that its interior must hold an even greater proportion of iron (65–70% by weight) than the Earth – perhaps also in a core.

A major discovery made by the Mariner 10 investigators was that Mercury possesses a significant magnetic field, one that apparently has a dipolar (north–south) nature like the Earth's. The Earth's magnetic field is generally held to be the product of electrical currents (created by a self-induced dynamo effect) flowing in the iron core. The observation of the Mercurian magnetic field (something of a surprise given the slow rotation of the planet) implies that Mercury also has a core. If this is the case than the radius of the core would be about 70–80% of the planet's radius (a core volume of about 50% of the planet's volume). The outer region of Mercury would then be composed of silicate rocks in analogy to the Earth's mantle.

Our understanding of Mercury at present indicates that it has an interior which generally resembles that of the Earth. The Mariner 10 data demonstrate that any resemblance to the Earth does not extend to the surface. The spacecraft confirmed that Mercury possesses essentially no atmosphere and the television images revealed a surface morphology that bears a striking resemblance to the Moon. Craters are the dominant surface feature, extending from the smallest size that could be resolved (about 100 metres) to multi-ring craters hundreds of kilometres across. Following the lunar convention, the latter are termed 'basins'. The most prominent of these is located near one of the two 'hot poles' – the one at 180° longitude – and has accordingly been named *Caloris*. The Mercurian basins differ significantly from those on the Moon in that they are not the sites of broad smooth maria (the result of the flooding of the lunar basins by basaltic lava). As a result the Mercurian landscape lacks the distinctive mare–highland dichotomy of the Moon. Otherwise, these two small planetary bodies can easily be mistaken for one another.

Mercury, like the Moon, is a relatively unevolved object. However, as has already been mentioned, there are important differences between the two planets; these will be discussed after first describing the major terrain types that are observed on Mercury.

Planetary geologists working with the Mariner 10 imaging data have mapped over a dozen different types of terrain. To the non-professional geologist, and even to someone who is accustomed to working with surface images of the Moon and Mars, many of the differences between the various terrain types are very subtle. Therefore, here only a few of the surface types will be discussed, beginning with what are believed to be the oldest and progressing to the youngest: (i) heavily crated terrain; (ii) intercrater plains; (iii) topography related to the Caloris Basin; (iv) smooth plains; and (v) young craters.

On the Moon and on Mars the earliest terrain types are those characterized by a high density of craters of all sizes. Such old *heavily cratered terrain* is also observed on Mercury where it differs, from the Moon in particular, in that around and between regions of heavy cratering there are significant areas of *ancient plains*. These plains on Mercury do not resemble the lunar maria – they are distributed among the heavily cratered regions rather than in broad basins and they are relatively highly cratered (whereas the lunar maria only record the later stages of meteoritic bombardment).

The relative ages of the heavily cratered terrain and the intercrater plains on Mercury are not straightforwardly determined on the basis of *superposition relations* (that is, which terrain lies on top of the other and is therefore younger): the plains do not clearly overlie the craters while, at the same time, ejecta blankets from the craters are not observed to cover the plains. The intercrater plains are well peppered by small craters which are judged to be *secondary* craters that are presumed to have been produced by the impact of debris excavated from *primary* craters in the heavily cratered terrain. If this interpretation is valid, then the plains must have already been emplaced at the time of formation of the heavily cratered terrain, or, perhaps, have been formed at about the same time. An episode of planet-wide surface melting, similar to that on the Moon, near the end of the massive early meteoritic bombardment might have created the plains. Alternatively, widespread volcanism, occurring simultaneously with the bombardment, might have been responsible.

The formation of the giant *Caloris Basin* was a key event in shaping the large-scale surface topography of Mercury. Not only are the effects of the event observed over a wide region centred near the equator at about 180° longitude, but, also, unusual hilly terrain at the antipodal region is thought to have been caused by the impact as a result of the focusing of seismic energy there. The Caloris Basin has a complex form. The most obvious characteristic is the series of mountain rings made up of smooth mountain blocks that rise a kilometre or so: these were presumably formed at the time of the impact. *Smooth plains* lying between the mountainous rings were evidently formed somewhat later, probably as a result of volcanic activity. In this regard these plains do resemble the lunar maria, although they are more heavily cratered. Such smooth plains are also located elsewhere on Mercury, including the floors of basins older than Caloris.

A variety of other terrain types, including some that have been labelled 'hummocky plains' and 'lineated terrain' are mapped in association with Caloris: their formation has not been linked with any key planet-wide process and they will, therefore, not be discussed further.

Turning now to the young craters, the Mariner 10 pictures show prominent rayed craters which are evidently the youngest features on the planet (because the ejecta material that makes up the rays overlies the other terrains). These craters are the result of meteoritic impacts that have continued at a low level ever since the main surface features of the planet were established. It should be remembered that, on a planet like Mercury where changes now occur with excruciating slowness, a 'young' crater may be more than a billion years old. Not all young craters have rays, since the 'gardening' of the surface by tiny meteorite impacts goes on all the time and eventually erases the superficial deposits that comprise the rays: a well-preserved crater form and a well-defined surrounding 'ejecta blanket' (a more substantial deposit than the rays) is sufficient to classify a crater as youthful.

The surface topography suggests that Mercury has undergone little surface evolution apart from meteoritic impact and limited volcanism. There is, however, more to tell. A careful examination of the Mariner 10 images reveals that there are, distributed across the face of the planet, quasi-linear features (*lineaments*) that are not associated with any one terrain type. These features, which are evidently the result of a disruption of the planet's crust (*tectonism*), consist of a variety of scarps (cliffs) and troughs (elongate depressions).

Typically the scarps (*rupes* in Latin) are between a half and one kilometre in height, and they may be several hundred kilometres long. The troughs are several hundred metres deep, typically about ten kilometres wide, and may run for more than a hundred kilometres. Over one hundred lineaments have been mapped on Mercury, so that useful statistical information on their distribution can be derived. However, the visibility of the lineaments is strongly dependent on the lighting angle in the pictures (the highlighting of a wall greatly aids in its visibility). It is, therefore, likely that many more remain to be discovered.

Attempts to understand the origin of the lineaments depend, firstly, upon the observed global distribution of such features and, secondly, upon the observed stratigraphic relationships of the lineaments with different terrain types; that is, upon the disentanglement of the relative chronology of the lineament-forming process and the formation of the different terrain types.

Information about the nature of the process in question is obtained from the distribution of the features: this will be discussed first. Careful mapping shows that the lineaments have a distinct tendency to be oriented northwest–southeast and, to a lesser extent, northeast–southwest. Towards the poles the orientation tends to be more north–south. What kind of tectonic process could have this result? Global contraction and/or expansion might be responsible: uniform shrinkage of the crust (perhaps due to internal cooling and contraction) would be expected to buckle the crust to form escarpments, while expansion would create trough-like cracks in the crust. However, interior changes that might lead to contraction or expansion are most likely to be spherically symmetrical and, therefore, would not create lineaments with the observed bias in orientation. So, other processes have to be considered.

Planet-wide tectonism could be produced by tidal despinning of a planet, a process that would not produce spherically symmetrical effects. In this concept Mercury was originally rotating relatively rapidly (as are most of the planets) and, subsequent to this early period near the time of formation, the rate of rotation was slowed as a result of tides created in the interior of Mercury by the Sun's gravitational pull. (Our own Moon has evidently undergone tidal despinning: the Moon's rotation has been slowed down, and the Moon has retreated from the Earth, until a 1:1 resonance between the rotational and orbital periods was reached at about 28 days.) In Mercury's case the rotation rate was evidently slowed until a 3:2 resonance was established. If the slowing down took place rapidly enough then crustal disruption, due to a flexing of the crust, could occur. Theoretical estimates suggest that the despinning of Mercury might have taken between 0.2 and 2 billion years. Theoretical analysis also suggests that east–west stresses would be induced that would be greater than those induced in a north–south direction. The calculated stresses are sufficient, for a reasonable assumption of crustal thickness, to fracture terrestrial rocks. Estimates of the expected orientation of consequent crustal fracturing are in reasonable agreement with the trends observed on Mercury, except at high latitudes where the lineaments tend towards a north–south alignment.

Because of the lack of agreement between theory and observation at high latitudes, it has been argued that tidal despinning on its own is an insufficient explanation for the lineaments. A rather more complicated scenario is postulated where despinning is accompanied by uniform global contraction so that high-latitude east–west trending fractures would be suppressed. As mentioned earlier, a uniform global compression of Mercury might have been caused by interior cooling such as might have taken place after an earlier heating episode caused by the gravitational effects of core formation. At this time the matter is by no means fully understood.

Having described some ideas about the nature of the process involved in the formation of the lineaments, the next question to address is *when* did they form? Here the examination of the stratigraphic record provides some answers. Relative ages of the lineaments are deduced from the manner in which they transect different terrain types. Thus, a scarp that is observed to cross a particular crater was clearly formed after the impact that created the crater. On the other hand, a scarp partially obliterated by a crater or other feature, must have been formed previously. The careful mapping of the lineaments reveals a history that may be summarized by saying that the postulated tidal despinning and global contraction occurred early during the period of heavy bombardment and intercrater plains formation. The Caloris impact was a later event as was the formation of the various smooth plains regions by localized volcanism.

In comparison to some other planets, Mercury has not had a very exciting history. However, Mercury has a special place at one end of the spectrum of planetary types. Although we may never be motivated to send astronauts to explore Mercury, future unmanned missions surely will be proposed since about half of the planet still remains to be seen, and there are numerous gaps in our understanding of the part that has been mapped. Furthermore, many of our notions about how Mercury evolved are without secure foundation.

This photomosaic and that facing the first page of this chapter were taken by Mariner 10 of Mercury as it approached and departed from the planet respectively. Until March 1974 Mercury was a planet of almost complete mystery. After Mariner 10 had made its first flyby on 29 March, the innermost planet finally entered the realm of the, at least partially, known. On that historic occasion (there were three flybys in all, 29 March 1974, 21 September 1974, and 16 March 1975) the closest approach was on the night side of Mercury in order to acquire optimum non-imaging science. However, although there was no opportunity to obtain very-high-resolution pictures, excellent global mosaics were acquired both on approach and departure. Each of these two mosaics comprises 18 frames taken over a 13 minute interval when Mariner 10 was about 200 000 km away from the planet. In the approach mosaic the view is mostly (about two-thirds) of the southern hemisphere, while the departure mosaic is equally biased to the north. Mercury is revealed to be remarkably lunar-like in some respects but quite different in others.

After the three Mariner 10 encounters with Mercury (including especially the second which was targeted to a closest approach on the dayside above the southern hemisphere to optimize imaging coverage) it was possible to produce high-quality topographic maps of about half of the planet. The available coverage is indicated here. After assembling the individual frames into photomosaics, airbrush maps were produced with strict cartographic control by artists of great experience: these maps are designed to depict surface topography only and do not show the effects of reflectivity variations (for example, no rays are seen in the maps). Taken together with the original pictures, the maps allow detailed geological mapping to be carried out. These maps also provide the non-specialist with an excellent grasp of the general nature of the surface of Mercury – better than can be obtained from the photomosaics. All nine airbrush maps are reproduced on the following pages together with selected video images. In this and all subsequent maps of Mercury 1 degree equals 43 km.

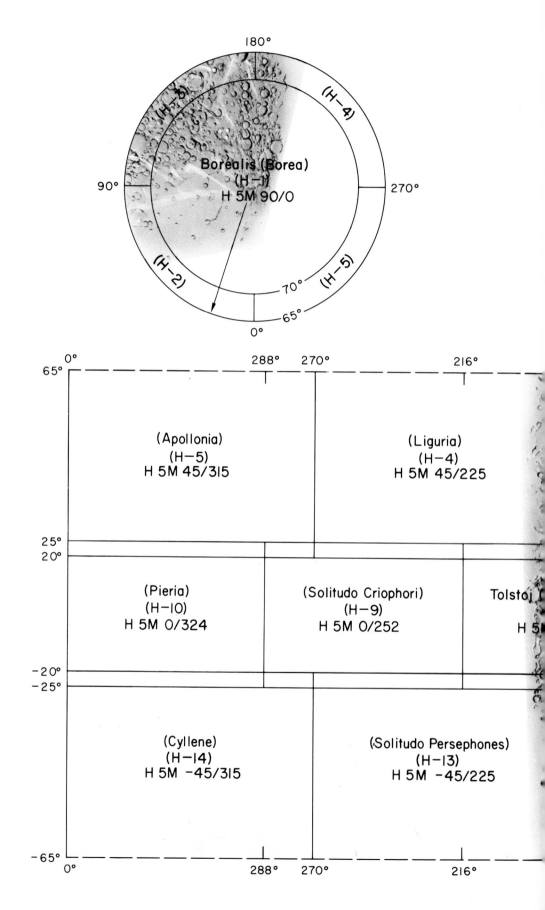

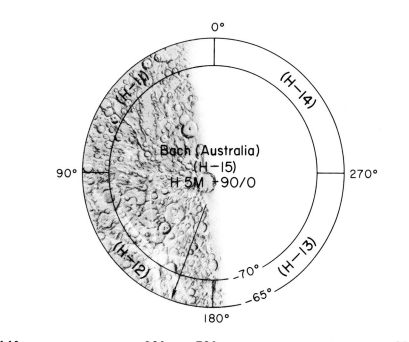

Bach (Australia)
(H-15)
H 5M -90/0

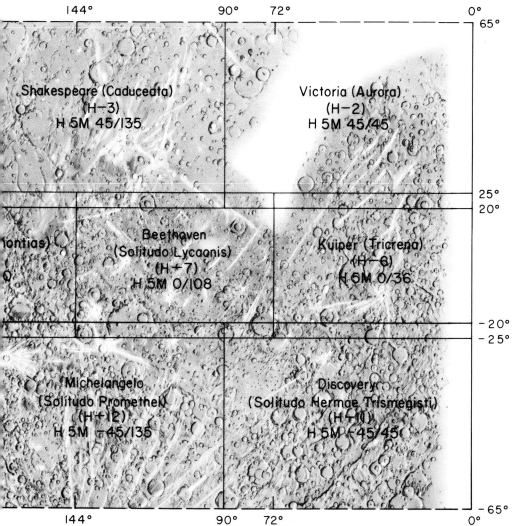

Shakespeare (Caduceata)
(H-3)
H 5M 45/135

Victoria (Aurora)
(H-2)
H 5M 45/45

Beethoven
(Solitudo Lycaonis)
(H+7)
H 5M 0/108

Kuiper (Tricrena)
(H-6)
H 5M 0/36

Michelangelo
(Solitudo Promethei)
(H+12)
H 5M -45/135

Discovery
(Solitudo Hermae Trismegisti)
(H+11)
H 5M -45/45

Mercury

The Borealis Area covers the north polar region. In the available coverage the terrain is approximately equally divided between heavily cratered terrain and younger, smooth plains which include the crater Goethe (about 300 km in diameter). Unlike the lunar case, the smooth plains are not significantly darker than the cratered terrain: their origin is therefore not as obviously due to basaltic volcanism as in the lunar case. An alternative that has been proposed is that the smooth plains are the result of blanketing by ejecta from impact events. However, the various characteristics of the smooth plains as observed within the available global coverage have led some geologists to the conclusion that at least some of the smooth plains units are indeed volcanic in origin.

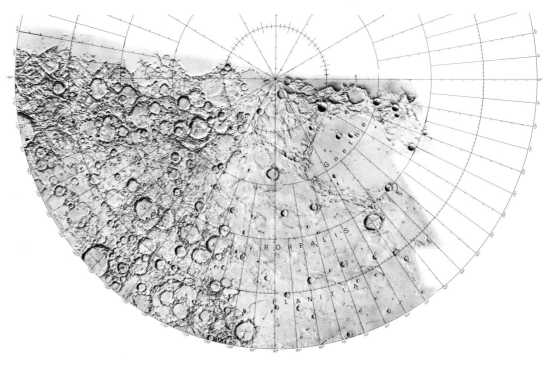

The 75 km diameter crater Saikaku (73°N, 177°W) appears at bottom right of this oblique view acquired near the morning terminator where shadows emphasize the ruggedness of the heavily cratered terrain in the Borealis Area. Saikaku and the other large craters to the north have rough flat floors. Between the craters are numerous smaller secondary craters formed in the intercrater plains. Reference to the map indicates that this is a region of very heavy cratering – probably the oldest surface unit on Mercury, the remains of the primitive crust formed while the meteoroidal bombardment was still very intense.

The Bach Area of
Mercury is centred on
the south pole of the
planet. One half of the
region was well imaged
on the second encounter
of the Mariner 10
spacecraft. This region
is generally lacking in
plains units.

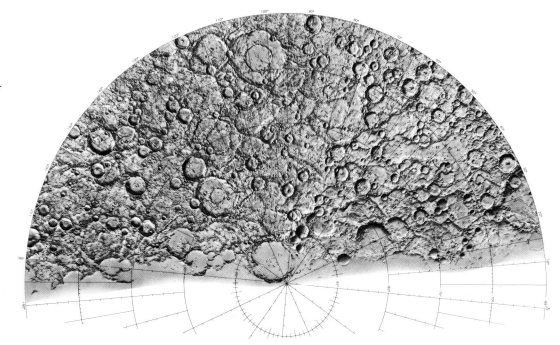

This photomosaic of an
enlarged view of part of
the Bach Area is another
carefully controlled
computer product in
which proper allowance
has been made for the
different lighting and
viewing angles of each
individual frame before
the overall picture was
assembled. Using the
accompanying map of
the Bach Area, the
craters Bach and
Wagner can be readily
identified: the
nomenclature adopted
by the International
Astronomical Union for
Mercury has led to a
juxtaposition of these
two famous names in a
manner that is
somewhat ironic in view
of their great distance in
the musical world.

The Shakespeare Quadrangle is perhaps the most representative of all the (arbitrary) map divisions of Mercury: it contains excellent examples of all the principal Mercurian terrain types. About half of the quadrangle consists of ancient heavily cratered terrain and intercrater plains. Smooth plains (e.g. Sobkou Planitia) and hummocky plains (e.g. Odin Planitia) also cover a substantial fraction of the quadrangle. To the west the great Caloris Basin, with its prominent encircling mountains, has led to surface modification over a large region. A typical linear feature, presumably formed early in Mercury's history as a result of massive crustal compression, is Heemskeck Rupes (Scarp) at about 25°N, 125°W.

This view of Sobkou Planitia (Plain) covers an area about 500 km on a side. To the left is the northwestern boundary of the plain and to the right are the large craters Bronte and Degas. This area is an excellent example of the widespread, localized smooth plains found on Mercury – a unit whose origin is controversial. One point of view is that these plains are analogous to the lunar 'light' plains (as contrasted to the dark *maria*) which are now known to have been created by ejecta from impact events. A second, more widely held, position is that, in many instances at least, the smooth plains are of volcanic origin: it is argued that the observed distribution of the smooth plains, including their frequent ponding within large craters, is much better explained by an internal origin, especially since obvious sources of impact ejecta are absent.

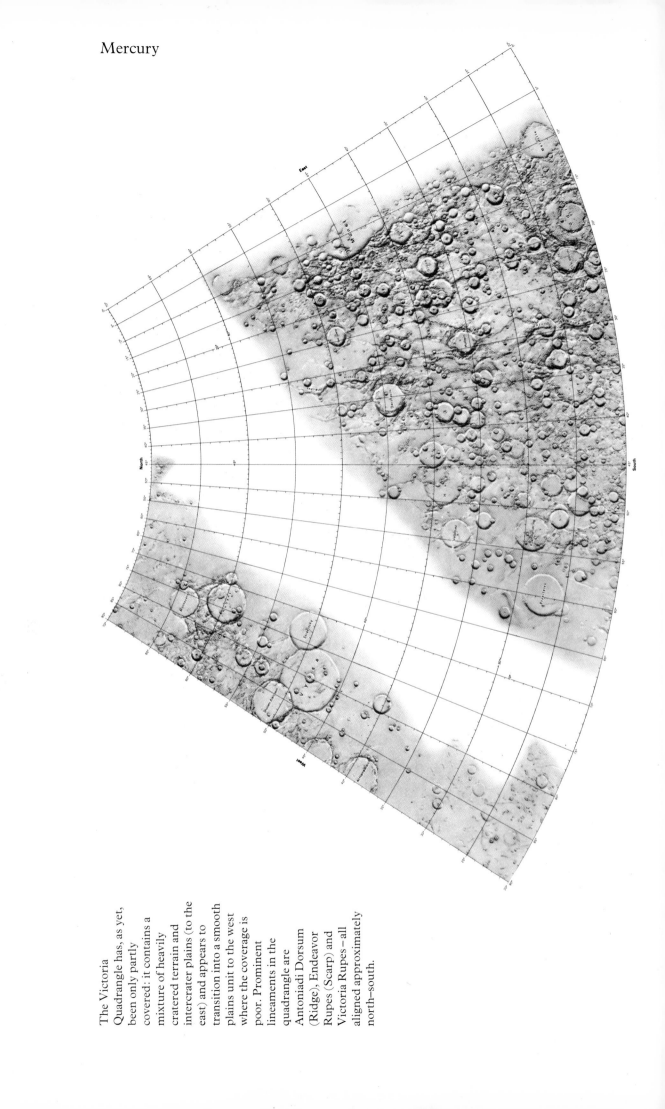

The *Victoria* Quadrangle has, as yet, been only partly covered: it contains a mixture of heavily cratered terrain and intercrater plains (to the east) and appears to transition into a smooth plains unit to the west where the coverage is poor. Prominent lineaments in the quadrangle are Antoniadi Dorsum (Ridge), Endeavor Rupes (Scarp) and Victoria Rupes – all aligned approximately north–south.

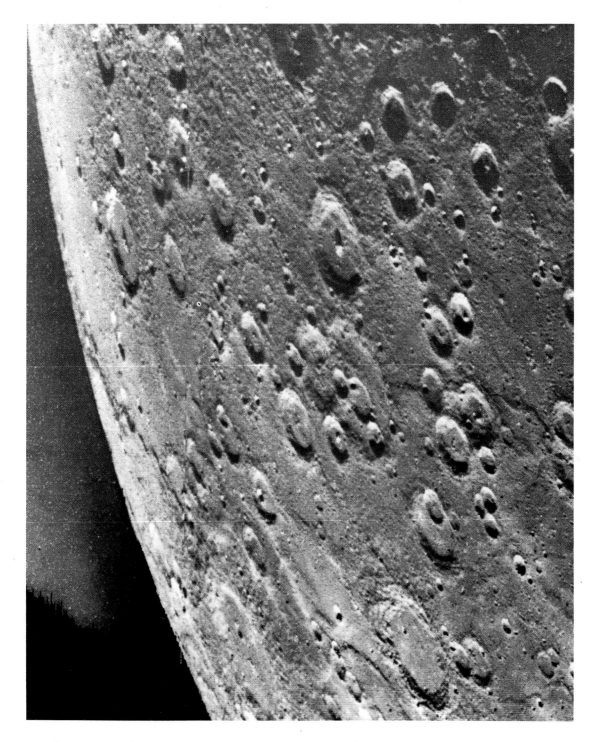

This oblique view of intercrater plains in the Victoria Quadrangle of Mercury stretches from about 30°N (bottom) to the limb which is at 50 to 55°N. Numerous large craters can be seen in the frame, many with central peaks. Between and around the large craters is terrain of a gently rolling nature described as intercrater plains. Geological analyses indicate that this terrain is about the same age as the heavily cratered units; perhaps the intercrater plains were formed as a result of an episode of planet-wide surface melting near the end of the early massive meteoritic bombardment.

Mercury

The Tolstoj Quadrangle includes the southern half of the great Caloris Basin which is located in the top middle of the quadrangle. While the available coverage shows a large percentage of heavily cratered terrain the region is notable for four large plains: Caloris, Odin, Budh and Tir. The 375 km diameter crater Tolstoj has been inundated, probably by lava, and is now also a region of (smooth) plains.

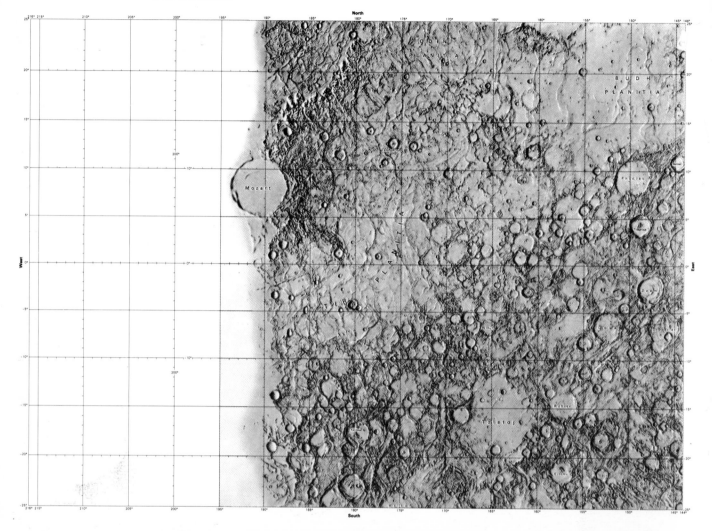

Large numbers of relatively fresh, bowl-shaped, secondary craters stud the intercrater plains that lie between the larger craters in the Tolstoj Quadrangle in this region to the northeast of the crater Mozart. The region is located just outside the ring of the Caloris Mountains which lie to the north. The frame covers an area of about 300 km by 250 km.

This view of the great Caloris Basin (so-named because it lies near the equator at 180°W – one of two 'hot poles' which alternately face the Sun when Mercury is at its closest distance from the Sun) covers about 1000 km in the east–west direction. In the northeast corner are the craters Van Eyck and Mansur, and in the southeast corner is the hummocky plains region Odin Planitia. The 1300 km diameter Caloris Basin, with its concentric rings of mountains, is comparable in many respects to the lunar Orientale Basin. The impact that created Caloris was a key event in Mercury's history since it served to modify the landscape over an enormous surrounding area. It also, apparently, led to the creation of unusual hilly terrain on the opposite side of Mercury as a result of the focusing of seismic waves there.

Mercury

The Beethoven
Quadrangle is primarily
occupied by heavily
cratered terrain and
accompanying
intercrater plains.
Beethoven and Raphael
are basin class impact
features within the
quadrangle while
Durer, Vivaldi, Wang
Meng and Boethius are
characteristic large
multi-ring craters.

This photomosaic is an
enlarged view of the
southeast region of the
accompanying map of
the Beethoven
Quadrangle. Surface
reflectivity (albedo)
variations have been
purposely suppressed in
the maps so that only
the topographic relief
remains. In the
approach and departure
mosaics a large amount
of albedo variation is
evident, particularly
rayed craters. This
computer-rectified
photomosaic of part of
the Beethoven
Quadrangle provides a
closer view of some
young Mercurian
craters. The rays were
created by the fall out of
debris excavated by
meteoritic impacts. The
three principal rayed
craters seen in this view
are located at about 8°S,
106°W, 13°S, 101°W,
and 7°S, 84°W.

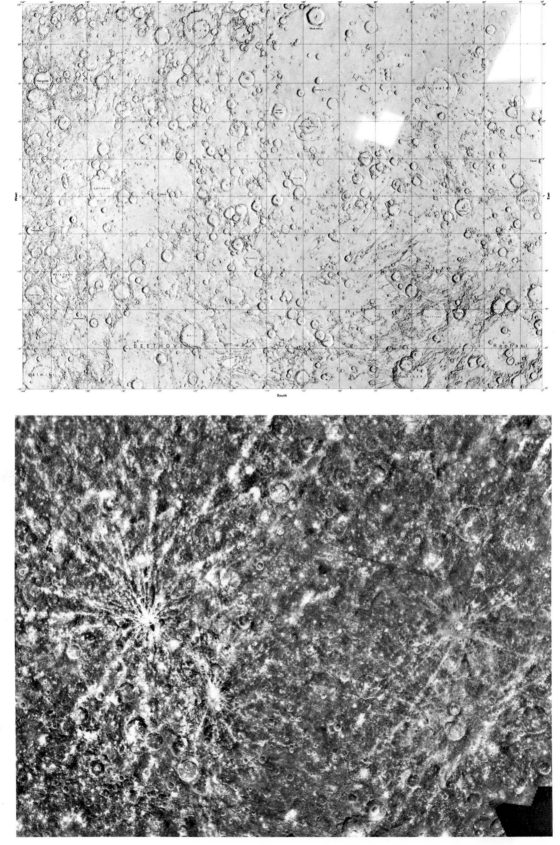

The Kuiper Quadrangle is dominated by ancient heavily cratered terrain, including several small (about 200 km diameter) multi-ring basins, e.g., Rodin, Homer, and Renoir. To the west the heavily cratered terrain merges into a region where the intercrater plains are more evident. Two of the large linear valleys in this quadrangle have been named in recognition of major U.S. planetary radar facilities. Goldstone Vallis at 15°S, 33°W is named after the Jet Propulsion Laboratory's facility in California (part of NASA's Deep Space Network which tracked Mariner 10 and other planetary spacecraft). Haystack Vallis at 5°N, 47°W is named after the Massachusetts Institute of Technology's facility located at Westford in Massachusetts. These troughs, found planet-wide, are linear in form and were evidently created early in Mercury's history.

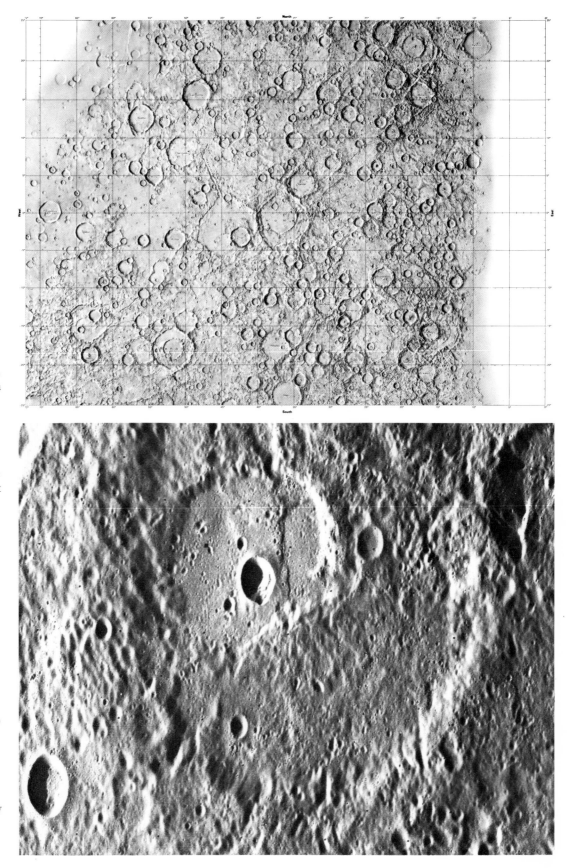

This Mariner 10 frame is centred at about 21°S, 28°W and covers an area of about 200 km by 110 km. An ancient, highly degraded (and unnamed) crater fills most of the frame. In the top left corner of the crater is a smaller (about 35 km diameter) degraded crater within which is a fresh bowl-shaped crater about 8 km across. The intermediate-sized crater has been deformed by tectonic forces creating a scarp which runs parallel to the crater's southern (bottom) rim for about 20 km then changes direction, moving north across the crater to the northern rim. The scarp crosses the rim and gradually peters out. Such scarps, and linear depressions, are found across the entire planet and are evidence of early crustal forces for which there is no analogy on the Moon.

39

Mercury

The Michaelangelo Quadrangle is dominated, like most of the others, by heavily cratered terrain and accompanying intercrater plains. Here also are found numerous lineaments, both scarps and troughs, that date back to an early period in Mercury's history when the crust buckled under the strains imposed by both planetary contraction and tidal despinning.

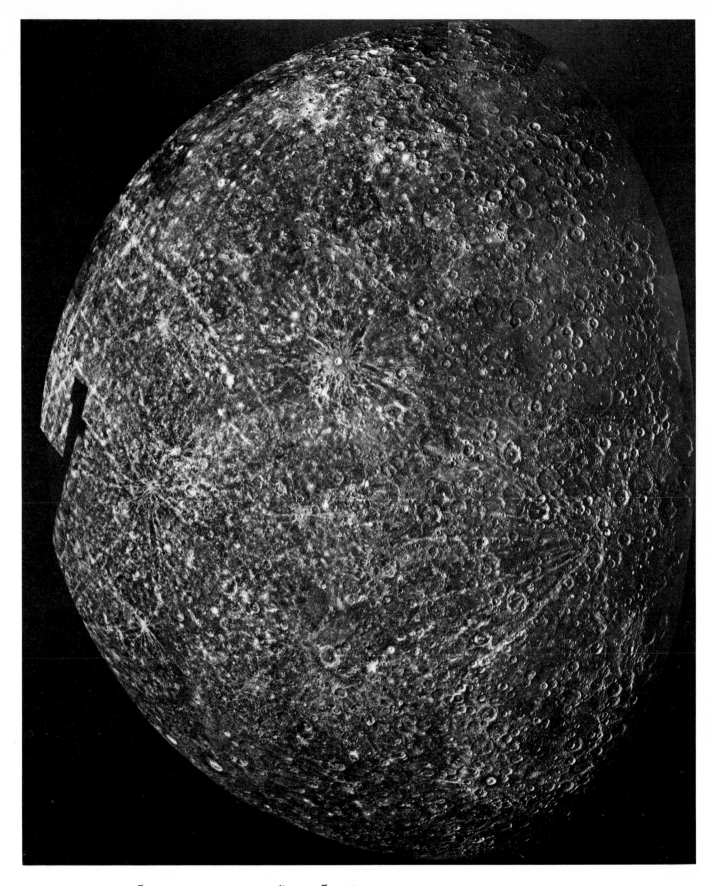

This exceptionally fine
photomosaic of part of
the Michaelangelo
Quadrangle was made
from numerous
individual television
images. Advanced
techniques were used
to make mosaics within
a computer rather than
by gluing together the
individual frames (the
most common
approach). As a result
the seams between the
frames are hard to
locate, creating the
effect of a single high-
resolution picture. The
high cost of computer
mosaicing (which
requires the interaction
of a skilled analyst)
precludes the universal
use of this technique.

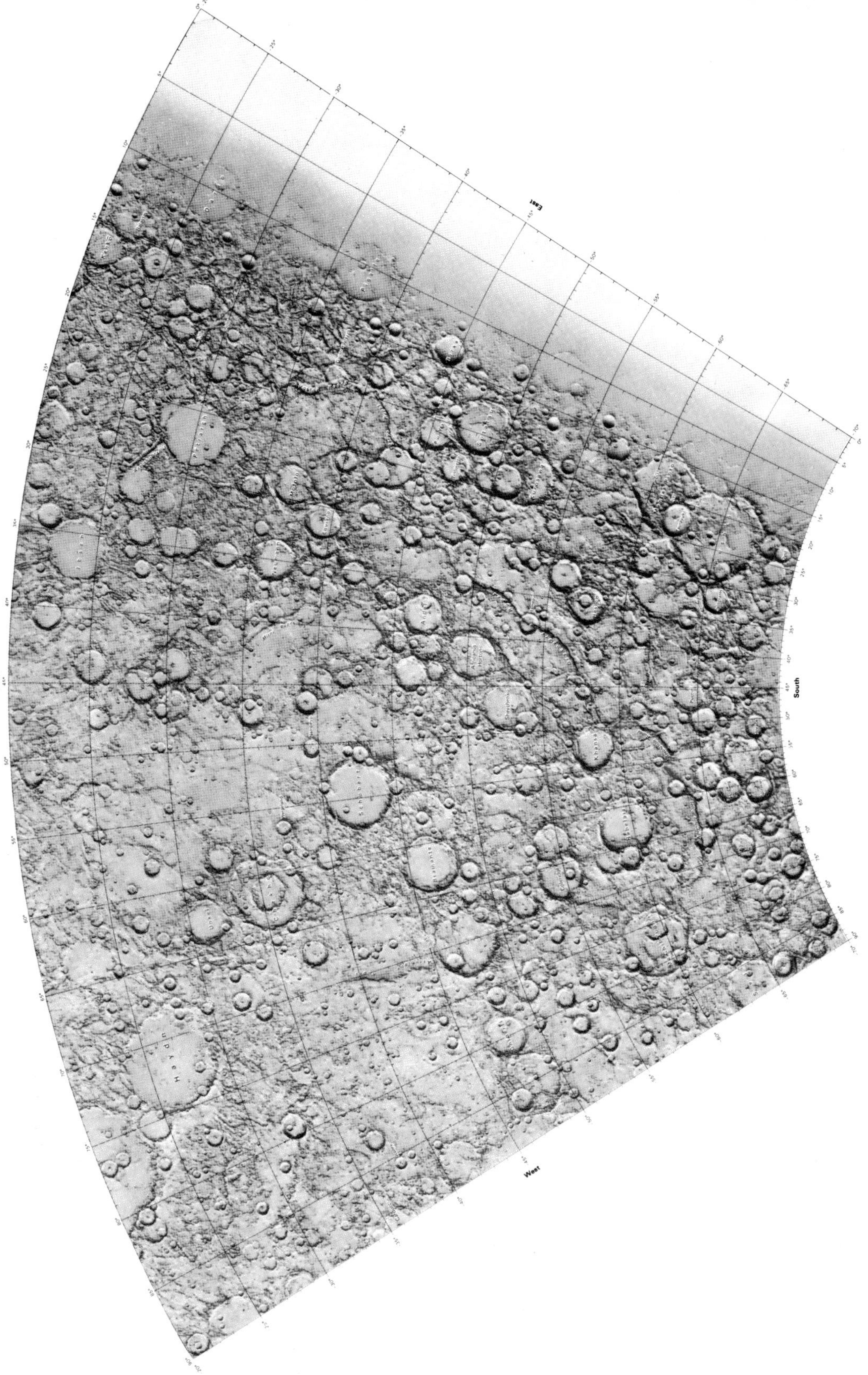

The Discovery
Quadrangle contains the
complete mixture of
Hermetian terrain types,
though ancient heavily
cratered terrain
dominates. Several of
the larger basins are
notable for their smooth
mare-like floors. This
quadrangle includes the
area approximately
antipodal to the Caloris
Basin, an area
characterized by
unusual hilly terrain,
possibly formed as a
result of seismic activity
associated with the
Caloris-forming event.
The quadrangle also
contains the best-known
example of a scarp on
Mercury: Discovery
Scarp (Discovery *Rupes*
on the map).

This frame at the Caloris antipode covers an area about 180 km on each side, centred at 31°S, 17°W. It shows a hilly terrain type, not observed elsewhere on Mercury, that has been hypothesized to have been formed as a result of focusing of seismic waves created by the Caloris Basin-forming impact on the antipodal region of the planet. On the Moon the antipode of the Imbrium Basin shows a comparably unique terrain.

The most notable scarp observed on Mercury thus far is Discovery Scarp, located near 55°S, 35°W. It is thought that such tectonic features as this were formed as a result of a combination of global expansion (due to interior heating) and tidal despinning (due to the gravitational interaction of Mercury and the Sun) early in the history of the planet. Discovery Scarp runs southwest–northeast for a distance of over 400 km, intersecting two pre-existing craters. Superposition relations such as these provide a means whereby a relative chronology can be attached to the different processes that are recorded on a planet's surface.

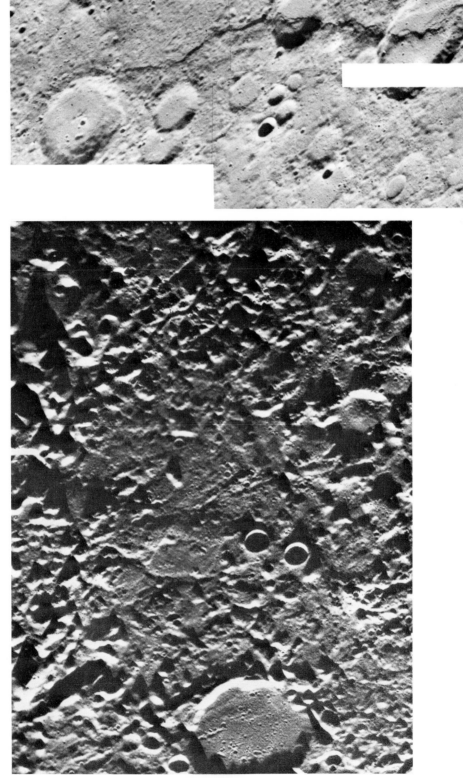

Venus

Venus is the closest planet to the Earth and, with the obvious exceptions of the Sun and Moon, it is easily the brightest object in our sky. Near inferior conjunction, when the two planets are separated by a mere 40 million kilometres, a modest telescope will reveal Venus as a crescent-shaped object with a brilliant, featureless aspect. Even the most powerful instruments and sensitive photographic plates show no markings on Venus which could be associated with continents, oceans or any of the surface features which abound on the other inner planets. Instead, only extremely subtle and ephemeral markings, and some 'scalloping' of the terminator which separates the day and night sides, have been reported by visual observers. Even the television cameras on the spacecraft Mariner 10, which observed Venus from a distance of 10 000 kilometres in 1973, were unable to detect significant contrasts over the brilliantly reflective disc of the planet when observing at visible wavelengths. The basic physical characteristics of the planet are summarized in the following table:

Mass 4.87×10^{27} g	Orbital period 224.7 days
Radius 6051 km	Obliquity $3°$
Density 5.2 g/cm³	Orbital eccentricity 0.007
Equatorial gravity 887 cm/s²	Mean distance from Sun 108.2×10^6 km
Rotation period 243 days (retrograde)	

It has long been realized that the high surface brightness of Venus, and the lack of visible features, must be due to a thick, uniform and permanent covering of clouds. The lemon-yellow colouration of the Venusian clouds was sufficiently different from their terrestrial counterparts to suggest to some that perhaps their composition was of dust, or some other material; but to most, the notion of a hot, tropical planet swathed in water clouds was irresistible. It seemed a logical expectation since, being an Earth-like planet with a thick atmosphere, Venus was presumably well endowed with oceans, which would more freely evaporate to generate clouds because of the enhanced sunfall at her reduced distance (0.72 AU) from the centre of the Solar System. Until relatively recently, the commonest picture of the surface of Venus in popular astronomy was one which resembled the primitive Earth: steaming coal forests, swamps and – sometimes – even dinosaur-like reptiles of great size. In many ways it is unfortunate that the concept of Venus as a new Earth had to die. However, the truth is sobering and full of significance for Earth dwellers.

In the 1950s, it became possible to measure the surface temperature of Venus for the first time. Microwave radiation – consisting of short radio waves with wavelengths typically measuring a few centimetres – is

emitted from the surface of Venus, passes unaffected through the cloud
layers, and can be measured on Earth. Its intensity is related to the
temperature of the emitting surface in a known way. The early results
for Venus showed temperatures of hundreds of degrees centigrade,
much too hot for free water or plant life.

Such high temperatures for the surface of Venus did not seem
reasonable; after all, the solar intensity is only twice that of the Earth,
and most of that is reflected away by the bright clouds, back into space.
In fact, Venus actually retains *less* solar power than Earth, and about
the same as distant Mars. Could the microwave measurements be in
error? It was mainly to answer this question that the first Venus
mission, Mariner 2, carried a small microwave radiometer. There was
no error – the temperature of the surface of Venus was found to be
more than 400°C. Even less doubt remained when the Soviet 'Venera'
series of spacecraft began their series of automated landings on the
planet's surface; not only was the temperature high by earthly
standards, but the pressure was found to be nearly a hundred Earth
atmospheres.

Under such conditions, the meteorology of the atmosphere and the
geological evolution of the surface may be expected to differ very
substantially from Earth's. Recently, the Soviets have mastered the
technique of softlanding their spacecraft on the surface of Venus, and
have obtained remarkable photographs of the terrain. These pictures
were obtained in natural light, showing the ability of a few percent of
the sunlight incident upon Venus to diffuse through the clouds to the
surface. As would, by now, be expected, the surface of Venus is
revealed as a sterile, scorched desert. Its appearance, at the Venera 9
landing site, is dominated by the boulders which appear strewn about
the landscape. However, there are significant differences between the
two landscapes, which are located about 2500 kilometres apart in the
equatorial regions of Venus. The Venera 10 panorama shows a flat,
rocky plain with a low outcropping of rock. It seems likely that Venera
9 landed on a slope (sensors on the spacecraft showed that it was tilted
at about 30° to the horizontal after settling on the surface) and that the
boulders are 'scree' or rubble from the break-up of the faces of the hill
upon which the spacecraft sits. Venera 10, on the other hand, sits on a
fairly flat, rocky plain. Both areas contain features which are consistent
with the ejecta and lava flows associated with volcanic activity. The
stones lying on the ground look much like an alluvial deposit, some of
them having dark bands and others a patchy appearance. They rest on a
mottled deposit which looks like crushed basaltic rock. The black
regions in this background are too dark to be shadows or even dark
minerals; they must be depressions or fissures between the rocks. Such

an interpretation is re-enforced by the very obvious sharp edges on some of the boulders, clearly the result of fracturing by some geological process. The fracturing could have been very ancient, since running water, large daily or seasonal temperature changes and wind erosion are not available to weather rocks on Venus as they do on Earth. On the other hand, some of the rocks do show evidence of erosion. Understanding the nature of this process is one of the major unsolved problems raised by the Venera pictures. Chemical erosion by acidic vapours in the atmosphere, or melting of volatile components of the rocks, are candidates under consideration. Such processes certainly must occur, but they probably lead to much slower weathering than takes place on Earth.

The level of illumination in the Venera pictures was higher than had been expected. Even with the Sun 60° above the horizon, it was thought that the thick clouds would prevent more than a trace of sunlight from reaching the ground; instead a light level which has been compared to that on Earth during a thunderstorm was discovered and the searchlights which the spacecraft carried were unnecessary. In 1978, radiometers on the Pioneer Venus probes measured the solar flux and found that $2\frac{1}{2}\%$ of the total falling on the planet actually reaches the surface without being absorbed. This is a large fraction considering the thickness of the cloud layers and shows, as do other measurements, how reflective the cloud droplets are on the whole. They diffuse the radiation thoroughly by scattering each photon dozens of times during its passage through the atmosphere, but do not absorb as strongly as terrestrial clouds would.

Other instruments on the Veneras, principally gamma-ray spectrometers, revealed the chemical composition of the surface rocks. At both sites, the abundances of the naturally radioactive elements uranium, thorium and potassium were consistent with a composition like terrestrial basalt. The density measurements, by gamma-ray backscattering, of 2.7–2.9 g/cm³, are consistent with this conclusion. Thus it seems that Venus formed in a manner similar to the Moon, Earth and Mars; by condensing from a molten protoplanet into shells, the outermost of which (the crust) is primarily composed of the most fusible material, the basalts.

The large-scale visible appearance of the surface of Venus is hidden from us for ever by the thick veil of clouds. However, much progress has been made in recent years in obtaining low-resolution radar pictures of Venus from the Earth, using the large dishes at Goldstone in California and at Arecibo in Puerto Rico to transmit and receive the pulses. In images obtained using this technique, regions of high radar reflectivity appear bright, indicative of variations in the composition or surface morphology. In other words, bright areas can appear so either because they are composed of material which is an intrinsically better reflector, or because the surface is smoother than surrounding darker areas, or because the surface is tilted to be more nearly normal to the incoming beam than the local horizontal. Three large, bright features dominate the map of the Venus globe at Earth-based resolution (about 85 km). Little is known of their geophysical nature, except that the surface must be quite rough; irregularities on the same scale as the wavelength of the radar beam (12.6 cm) or greater is implied by the backscattering properties of these regions.

Radar mapping can be accomplished much more readily by a small radar on a spacecraft. The first to attempt this was the Pioneer Venus Orbiter, part of a multiple mission to Venus by NASA in 1978 which is described in more detail later. Because it was placed in a high inclination orbit (that is, passing over the polar regions) it was able to map a much greater fraction of Venus's surface, as well as obtaining better spatial resolution and relative height information. Maps were slowly built up strip by strip as the orbit precessed around Venus, taking a Venus year of 243 days to complete.

The Pioneer Venus global altimetry coverage has provided a breakthrough in our knowledge of the solid body hidden beneath the clouds. Even though the horizontal resolution of the radar data (50 km) is crude in comparison with the imaging resolution available for other Solar System bodies, the radar map does immediately tell us generally what Venus is like. We find a planet that is characterized primarily by smoothly rolling plains (about 70% of the surface area), with clearly distinguished highland (about 10%) and lowland (about 20%) regions. The highland areas are uniquely Venusian, as may be judged by discussing the two most notable of the half dozen elevated regions. One has been named *Ishtar Terra* and is located at about 70°N. Ishtar covers an area comparable to Australia and rises steeply from the surrounding plains. The western part of Ishtar is a high plateau (3–4 km above the mean radius of Venus) bordered by tall mountains that reach a further 3 km in altitude. In the middle of Ishtar stand the *Maxwell Montes,* mountains that, in rising to 11 km above the mean, would tower above Everest. A large roughly circular depression near the centre of Maxwell might well be a caldera.

Stretching for about 10 000 km along and south of the equator, *Aphrodite Terra* is the other most prominent highland region, one that covers an area about equal to that of Africa. Aphrodite is even more rough and complex than Ishtar and contains an enormous (about 2400 km in diameter), roughly circular feature without obvious analogue, and several gigantic trenches – kilometres deep, hundreds of kilometres wide, and over 1000 km in length. These linear depressions are likely to prove as dramatic as the *Valles Marineris* on Mars when, in the future, they are captured in high-resolution radar images.

The rolling plains that cover most of Venus are relatively undramatic when observed at the low-resolution presently available and when compared to the rough highland areas. Nevertheless, they clearly contain a diversity of features, including circular structures that range to over 1000 km in diameter – features that could be impact craters or may have an igneous origin.

It is clear that Venus has had a complex history and, until high-resolution images of the surface are available, we will remain tantalized. Are the highland regions the Venus analogue of terrestrial continents or are they gigantic volcanic piles more like the surface of Mars? Do the rolling plains date back into the distant past or are they relatively recent in origin? Will we see evidence of current or past plate tectonics?

Not all planetary scientists are content to wait for the next big step in the exploration of the surface and interior of Venus and are, instead, bringing together all of the available geophysical (e.g. gravity field and altimetry) and geochemical (e.g. atmospheric composition pertaining to the abundance of radioactive elements in the crust) data to try to put together a picture of what Venus is like on a global scale. For instance, the fact that the distribution of surface altitudes on Venus is strongly monomodal while that on Earth is distinctly bimodal suggests to some that plate tectonics is not, or is no longer, an important process in shaping the surface of Venus. Put more simply, the Earth is seen as being covered with large plates which are separated by ridges and troughs on the ocean bottoms and which can move slowly in relation to each other. In contrast, Venus is more uniformly flat with a few eruptions of high ground suggesting that this surface may have evolved to become a single plate. This could have happened if the crust of Venus was more mobile than Earth's, either because it once contained a lot of water (which reduces the temperature required for partial melting of the mantle), or perhaps because of critical phase differences in silicates (bearing in mind that Venus is slightly less dense than the Earth). We must also consider what effect the slower rotation rate and absence of a large moon would have on the mobility of the crust of Venus relative to that of Earth.

In contrast with these arguments we have the fact that high features on Venus, although few, are very high when they do occur. This suggests that either the outermost layers on Venus are no longer very plastic (as if they had once contained a lot of water which subsequently was lost, first into the atmosphere and then to space) or that the topography is very young, and continuously renewed. For the former case to apply the rocks on Venus would have to be very dry by terrestrial standards – the Earth's crust could not support mountains as large as Maxwell for long at Venus's temperature. However, high features on Earth are supported principally by *isostatic* processes (forces exerted between the plates) while those on Venus are probably supported by convection. The later process, which occurs on Earth as a secondary phenomenon (for example, in producing the Hawaiian islands), involves plumes of hot, low-density material rising slowly through the crust, producing 'hot spots' and dynamically maintaining a surface protrusion. The shapes of the mountains on Venus resemble convective features on Earth. Maxwell, for example, looks like an ancient segmented volcano. The so-called beta region, on the other hand, appears to be a basaltic shield volcano with a radiating pattern of relatively fresh lava flows – a much younger feature than Maxwell. Indeed, there is a reasonable chance that the beta volcanoes are still erupting, perhaps contributing copious amounts of carbon dioxide, sulphur dioxide and water vapour to the atmosphere. Such suggestions point again to the complex coupling between atmospheric and surface processes and to the delicate balance that has taken near-twin planets on diverse evolutionary paths.

It was noted earlier that observations of Venus at visible wavelengths, whether made by an observer at a telescope or by a camera on a spacecraft, show little or no detail in the clouds which seem to form a uniform blanket over the whole planet. However, it has been known since the 1920s that photographs taken through an ultraviolet filter show blotchy features in the cloud. Under the best observing conditions, these markings often show a characteristic shape, like a letter 'Y' laid sideways. It is still not entirely clear why Venus shows contrasts in the ultraviolet and not the visible spectrum – probably some uv-absorbing material such as gaseous SO_2 or elemental sulphur is non-uniformly dispersed through the clouds. SO_2 is definitely present in ultraviolet spectra observed from the Pioneer Venus orbiter, but its spectrum does not match that of Venus at all wavelengths, and some other material, such as sulphur, which absorbs uv but not visible radiation, must be contributing also. Polarimetric and spectroscopic observations from the Earth have identified the main component of the uppermost cloud as concentrated sulphuric acid droplets; it is not surprising, therefore, to find small amounts of S and SO_2 also present. Whatever their origin, the uv markings are of immense interest, because their structure, movements, and evolution provide clues to the cloud structure and meteorological activity on Venus. Indeed, it was apparent by the early 1960s that the Y-feature appeared to rotate around the planet in a period of only 4 to 5 days. This implied wind velocities of 100 metres per second in the cloud tops, surprising for such a slowly rotating planet. The solid surface of Venus rotates at only about 4 metres per second, or once every 243 days. It was logical that Mariner 10 (in 1974) and the Venera 9 and 10 and Pioneer orbiters (in 1976 and 1978 respectively) should image the Venusian clouds in the ultraviolet to confirm this finding and look for further details in the structure and motions. As soon as the first pictures reached the Earth it was immediately obvious that the ultraviolet features which the astronomers had glimpsed are real, and furthermore that they exist on the smallest spatial scales viewed to date (a few kilometres).

Comparison of the Y-shaped markings as inferred from a composite of Earth-based observations in 1966, and as reconstructed from a mosaic of Mariner 10 ultraviolet photographs taken during the period from 5–12 February, 1974 shows a striking similarity. What atmospheric processes on Venus give rise to this behaviour? The meteorology of Venus is as mysterious as its surface geology. It is apparent that a large propagating wave, having a wavelength equal to the equatorial circumference of the planet, is present, and somehow manifests itself in the appearance of the clouds in the ultraviolet. In order to synthesize the Y-shaped feature, the wave motion must be fairly complex. It has been suggested, by analogy to terrestrial phenomena, that the superposition of two planetary-scale waves, one dominant near the equator and the other at the mid-equatorial latitudes, may account for the Y. Careful measurements of the propagation velocities of small scale features (thought to move with the winds) and that of the large Y (representing the phase velocity of the wave or waves, superimposed on the wind velocity) have confirmed that the bulk velocity of the atmosphere near the equator is 100 metres per second, a result which is borne out by Earth-based measurements of the Doppler shifting of spectral lines, in the zonal direction (parallel to the equator). The global-scale waves propagate upstream at about 20–30 metres per second. Both of these velocities are much larger than the apparent velocity of the Sun with respect to an observer on the surface. Probe and remote sounding measurements show that the rate at which the atmosphere circulates around the equatorial regions varies considerably with height, its maximum speed occurring near the cloud tops where the trackable ultraviolet markings are thought to originate. Above the clouds the 100 metre per second winds are decelerated sharply by the pressure gradient which results from the temperature distribution at those levels. Rather remarkably, Pioneer discovered that the air temperature is some 15–20 °C *warmer* at the pole than the equator from 70–95 km above the surface. Dynamical models imply that this type of gradient is sufficient to arrest the zonal winds completely by

85–90 km altitude. Below the clouds, the winds fall gradually in velocity as the atmosphere becomes denser; doppler tracking of the Pioneer probes shows a wind speed of less than 10 metres per second at the 10 km altitude level and close to zero at the surface. All of the zonal winds are westward (in the same direction as the rotation of the planet), which means that angular momentum is being delivered to the atmosphere by the solid body of the planet and transported upwards. Alternatively, it has been suggested that the Sun exerts a torque on the atmosphere and so supplies external angular momentum. This it certainly will do, since the density of the atmosphere is non-uniformly distributed with solar longitude (local time of day) because of thermal tides induced by solar heating. In fact, the semidiurnal component of the tide, on which the torque principally is exerted, has been observed to be unexpectedly large, relative to the diurnal component, on Venus, which favours this mechanism. Whether the effect is large enough to accelerate the atmosphere to the speeds observed is a subject of intense debate. There is even a possibility that the slow retrograde rotation of the planet itself may have been established, over geological time, by the torque which the atmosphere exerts on the planet – the reverse of the earlier theory.

Less dramatic than the zonal winds, but of even greater significance for the general circulation of Venus, is the observed migration of the ultraviolet markings away from the equator towards both poles (i.e. in the *meridional* direction) at speeds of less than 10 metres per second. The general impression is of two gigantic circulation cells, one to each hemisphere, in which (heated) air rises at the equator and (cooled) air descends at the poles, travelling more or less horizontally in between, polewards above the clouds and equatorwards below. Such a flow is the characteristic of a Hadley cell, the simplest circulation regime which can occur in a planetary atmosphere. This kind of structure was proposed for the Earth's atmosphere by Hadley as long ago as 1735. It seemed logical to him that rising air at the warm equatorial regions and falling air at the cool poles, would lead to pole-to-equator flow near the surface and to motion in the opposite direction at higher levels. This does in fact happen, but in our own atmosphere this simple structure is greatly modified by the development of 'baroclinic instabilities' in the motion. In simple terms this means that the smooth flow of the Hadley regime tends to break down under the influence of the Earth's spin. The net result on Earth is that Hadley's cell extends from the equator only to mid-latitudes, with other, smaller, cells taking over the transport nearer the Pole. On Venus, it appears that the basic Hadley configuration may exist in a less modified state.

Dramatic evidence for this type of circulation was obtained by infrared measurements from the Pioneer Venus orbiter, which provided the first observations from above the polar regions. Greatly enhanced amounts of infrared flux were found to be emerging poleward of 80°N latitude, in a localized, elliptical region which evidently is a clearing in the cloud cover forced by the descending air in the return branch of the Hadley cell. This clarification of the circulation regime existing near the cloud tops also gives us some clues to other fundamental questions about Venus, namely its near-uniform cloud cover and its high surface temperature. Probably, the air is rising slowly everywhere on Venus and descending only near the poles, thus supporting a single planet-wide cloud. This is in marked contrast to all of the other cloudy planets we know well, where the tendency is for more localized circulation cells to form with clear regions in the descending parts. The basic difference between Venus and Earth, Mars, Jupiter and Saturn in this respect is the slow rotation of Venus, which allows the monocellular (Hadley) regime to be stable – notwithstanding the rapid zonal winds superimposed upon it. The planet-wide, thick, blanket of cloud plays a key role in trapping heat radiation in the lower atmosphere and raising the surface temperature.

Other features of great interest and which have been observed in the atmosphere of Venus are the various planetary-scale wave motions. We have already mentioned the large Y feature whose detection from the surface of the Earth began the now burgeoning interest in Venus's meteorology. Some of the other important wave phenomena on Venus are described in the following paragraphs.

The *circumequatorial belts* are very narrow (less than about 50 km in width) features, of variable length (of the order of thousands of kilometres) and transient appearance. As many as five have been seen at once, evenly spaced by about 500 km and always aligned parallel to the equator. They appear in 1 or 2 hours and propagate, always in a southerly direction, for about $\frac{1}{2}$ to $1\frac{1}{2}$ days at about 20 metres per second. A time-lapsed sequence of photographs shows a segment of a belt travelling along the subsolar meridian. The most satisfactory explanation for the belts is that they are some form of gravity wave. Gravity waves are 'resonances' in the atmosphere, caused by density variations propagating as waves under the influence of gravity as the restoring force. They are common in the Earth's atmosphere, and indeed they have been observed in satellite pictures of terrestrial clouds because temperature fluctuations, associated with the density waves, lead to condensation in the thermal troughs. Something similar may be happening on Venus. It is far from clear, however, what is exciting the Venusian waves. It could be turbulence in the strongly heated subsolar region, or perhaps some lower atmospheric wave propagating upwards. Still, it is difficult to explain why the waves always seem to travel from the north to the south.

The *bow-like waves* were named for their shape (like that of a bow, as in archery). However, it turns out that they probably have something in common with the bow waves associated with the passage of the bow of a ship across water. What appears to be happening on Venus is powerful 'boiling' of the atmosphere at the region directly below the Sun, i.e. at local noon. The rising of the heated air, visible as convection cells in the images, interferes with the smooth, high velocity flow of the upper atmosphere and generates 'ripples' in the clouds. However, this explanation is even more limited than its obvious oversimplification would imply, since the waves travel downstream behind the subsolar zone, whereas oceanographic bow waves remain fixed with respect to the disturbance producing them.

The *subsolar disturbance* itself appears in the photographs as a pattern of cellular features, irresistibly suggestive of strong convection. In fact, the appearance of two types of cells (light with dark centres, and vice versa) on Venus would seem to correlate with the occurrence of open (descending at the centre) and closed (ascending at the centre) cells in convective regions of the Earth's atmosphere. The terrestrial cells have a characteristic ratio of diameter to depth, and occur only in regions of low wind shear. If these characteristics extend to Venus, then the convective region is about 15 km deep, and the wind shear does not exceed 2 or 3 metres per second per kilometre. This in turn implies zonal (east–west) velocities of 50 metres per second or more at the bottom of the convective layer.

Two of the most prominent and puzzling of Venusian atmospheric phenomena are observed only in infrared images of the planet. These use the thermal emission from the planet as a source, and so cover the dark or unfavourably illuminated portions of the planet as well as the day side. They also reveal the temperature and vertical structure of the phenomena which they observe. The *circumpolar collar* is a current of very cold air which surrounds the pole at a radial distance of about 2500 km. It is nearly 1000 km across, but only about 10 km thick in the vertical direction. Temperatures inside the collar are about $30\,^\circ$C colder than at the same altitude outside, so the feature generates pressure differences which would cause it to dissipate rapidly were it not continually forced by some unknown mechanism.

Inside the collar lies the region of reduced cloud cover caused by the descending branch of the Hadley cell. Because of the zonal momentum transported from lower latitudes, the descending air is also rotating rapidly, forming a *polar vortex* analogous to the eye of a terrestrial hurricane or whirlpool but much larger and more permanent. Interestingly, however, the eye of the Venus polar vortex is not circular but elongated, and with brightness maxima (presumably corresponding to maximum downward flow) at either end. This gives the pole a 'dumbell' appearance and has led to the name *polar dipole* for the feature. The dipole rotates about the pole every $2\frac{1}{2}$ to 3 Earth days, i.e. with about twice the angular velocity of the equatorial cloud markings. If angular momentum were being conserved by a parcel of air as it migrated from equator to pole the dipole might be expected to rotate five or six times faster. In fact, the uv markings are observed to keep a roughly constant angular velocity (solid body rotation) from the equator to at least 60° latitude, presumably accelerating poleward of this. This assumes, of course, that the rotation of the dipole represents the actual speed of mass motions around the pole and not simply the phase speed of a wavelike disturbance superimposed on the polar vortex.

What is the overall picture of the atmosphere of Venus which is emerging from these observations? A mixture of the recently discovered facts with a certain amount of speculation yields the following overview.

Heat radiation from the relatively nearby Sun is mostly reflected back to space; only about 15% is absorbed in the thick layers of cloud at heights from 30 to 60 km above the surface. A small amount (around $2\frac{1}{2}$%) of the solar energy reaches the surface itself; this is enough to maintain the high temperature of the lower atmosphere because the thick atmosphere and near-total cloud cover trap outgoing heat radiation (the greenhouse effect). The heating effect in the clouds themselves causes convective rising motion; the warm air propagates towards the poles, cools, and descends. It is the presence of these two large cells in the middle atmosphere (additional cells have been postulated above and below) which accounts for the gross appearance of Venus, and also for its ability to sustain a large greenhouse effect. Both can be attributed to the absence of significant amounts of large-scale downwelling, except over limited regions at the poles. In turn, the ability of Venus to sustain a relatively simple Hadley-type circulation is probably attributable to the very slow rotation rate of its solid body. The meridional cells are twisted into spirals by having a rapid zonal flow superimposed on them. No comprehensive theory of the forcing of the zonal winds exists yet; in particular it is not yet clear whether the angular momentum involved originates in frictional interaction with the surface, or gravitational interaction with the Sun. However, it is now known that they are decelerated again high above the clouds by the pressure gradients associated with a relatively warm polar cap at these heights.

The clouds themselves contain sulphuric acid, hydrochloric acid and water, and probably a good many other things besides. The ultraviolet contrasts are due, in part at least, to varying amounts of sulphur dioxide gas mixed in the clouds. The equatorial Y-shaped feature, the circumpolar collar and the double 'eye' in the polar vortex are large planetary-scale waves unlike anything found on Earth. Their detailed appearance is changeable, but the Y and the dipole rotate while the coldest part of the collar stays at the same local time of day. The latter could be caused by the subsolar disturbance propagating polewards in the meridional cell; the appearance of the collar itself at a latitude of 70° or so probably marks a discontinuity in the *zonal* flow, perhaps the transition between a regime which conserves angular momentum and one which tends more to solid body rotation, as described above. The double structure at the pole is caused by a wave motion which modifies the vortex at the centre of the spinning Hadley cell. Such an arrangement may be necessary to transport the large amounts of angular momentum, which arrive at the pole, downwards more efficiently (for a given mass of descending gas) than a simple vortex could.

Interestingly, the thermal tide on Venus around the equatorial regions also has two maxima and two minima. (The thermal tide is simply the diurnal increase and decrease of temperature caused by the rising and setting of the Sun.) This does not seem to be connected with the polar dipole, since the two regions are separated by a narrow latitude band apparently free of planetary-scale waves, as well as by the predominantly wavenumber-one collar. The Earth's atmosphere has a wavenumber-two component superposed on the familiar early-afternoon maximum to postmidnight minimum cycle, but this does not explain immediately why this component should dominate on Venus. Some property of Venus's atmosphere apparently causes it to resonate and amplify the wavenumber-two tide, even though the forcing is predominantly wavenumber-one. To understand this kind of behaviour generally requires detailed numerical modelling of the atmosphere, by solving the relevant time-dependent equations on a large computer. In principle, the physics of the atmosphere can be represented by the model in such a way that the observed behaviour is reproduced. Models of Earth's atmosphere are approaching a level of sophistication where they might usefully be employed to forecast the future. For Venus, we would be satisfied in the short term to be able to reproduce the equatorial tides, the polar collar, and the rotating dipole.

A Pioneer Venus image of the planet at full phase, obtained through an ultraviolet filter. Note how the contrast of features in the cloud cover is enhanced compared to the visible image facing the first page of this chapter.

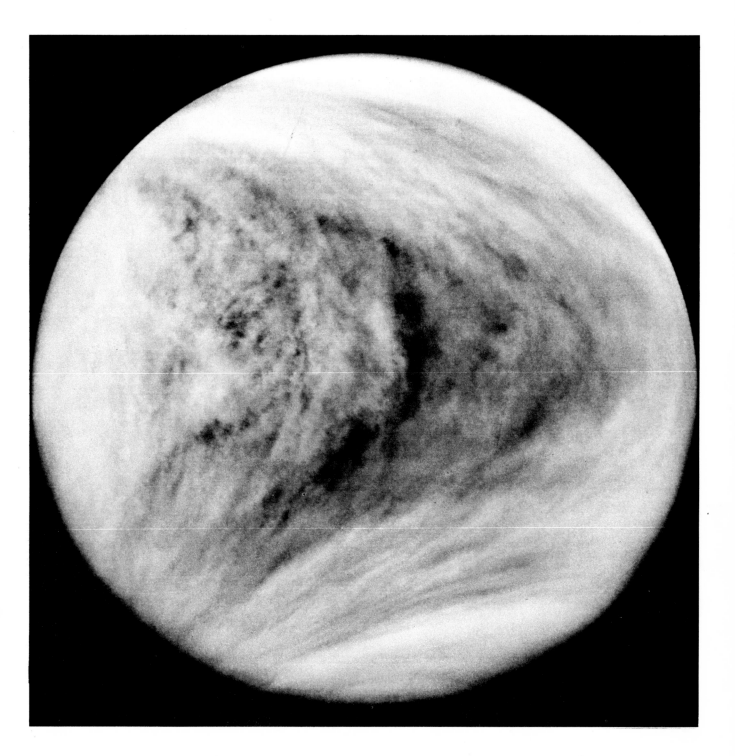

This was the first-ever television image of the surface of Venus, obtained by the Soviet softlanding spacecraft Venera 9 on 22 October, 1975. The camera itself was a line-scanning device, designed to survive the tremendous temperatures and pressures at which it had to operate by viewing the outside world from inside an hermetically sealed, thermally insulated vessel via a scanning mirror. The articulation of the scanning mirror allowed for rotation about a 'panning' axis, as well as a rocking motion in the plane parallel to the plane containing the axis, thus building up a $40° \times 180°$ picture of the surface during a 20–25 minute 'exposure'. Boulders, typically tens of centimetres in size, are seen strewn about a rocky landscape. This seems to have been produced by the break-up of the hillside on which the spacecraft sits, and fairly recently in geological time. This would be consistent with volcanic activity in this area. The round object in the foreground is part of the lander; the T-shaped device is the gamma-ray densitometer. The horizon, some 50 m distant, can be seen in the top right corner; it is not visible at the other side of the picture because of the 30° tilt of the spacecraft. The top picture is distorted by the unusual scan geometry of the camera. Below, the data have been computer-rectified so that the view appears as it would to an observer at the landing site.

Venera 10 was identical to 9 and landed 3 days later, taking this picture (top) of the surface about 2000 km away from its sister craft. This spacecraft landed in a more upright position than the other, and so the horizon appears more distant (perhaps 500 m). It is sitting on a large platform of rock, light in colour, which resembles terrestrial basalt. Some of the rocks are fractured, and others eroded, by processes which are not understood at present. Material which looks like soil is present between the boulders and in general the site looks geologically older (i.e. more eroded) than near Venera 9. Again, the lower picture is a rectified version.

Venus

Venus as seen by radar from the Earth. This picture was built up using the large (64 m) antenna at Goldstone, California, as transmitter and receiver; the same dish is also used for tracking spacecraft. The bright features on Venus are difficult to interpret uniquely in this kind of data, as several different mechanisms can give enhanced radar reflectivity. For example, a hard, flat surface reflects better than a powdered one and an Earth-facing slope reflects better than an oblique one.

About one day after its closest approach to Venus, Mariner 10 obtained this ultraviolet image of Venus which shows a variety of detail in the cloud structure. Note particularly the south polar ring and the spiral streaks. These patterns are suggestive of a spiralling motion towards the pole and this is confirmed by tracking individual features. This image was obtained at a distance of 720 000 km through a 3550Å filter.

Venus

A topographic map of the surface of Venus obtained from Pioneer Venus radar altimeter data. Most of the surface is a relatively smooth plain, here coloured in shades of blue. Occasional depressions and basins (puce and purple) descend up to 3 km below the mean level of the plain and cover about 16% of the surface – far less than their terrestrial equivalents. The highlands (yellow, orange and red) form two large and several smaller 'continents'. The

most northerly (Ishtar) is about the size of Australia and rises an average of about 4 km above the mean surface level. The high feature at the centre of the Maxwell Montes region is a mountain taller than Everest (11 vs 9½ km). It is believed to be an old volcano, probably now inactive; the caldera at the top and possible evidence of old lava flows can be seen. The equatorial continent (Aphrodite) is comparable in size to Africa and is more variable in

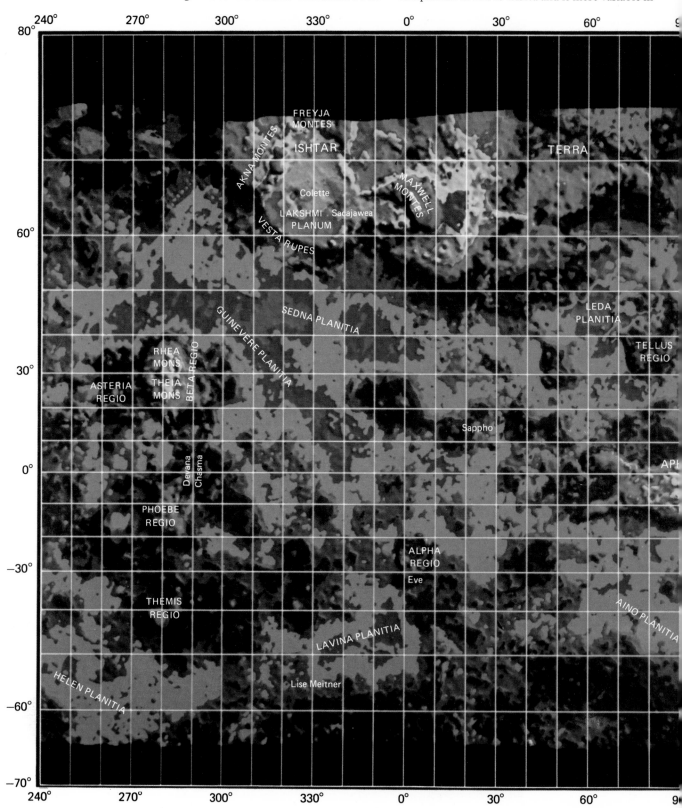

height than Ishtar. Its resemblance in shape to a scorpion has been noted. The eastern side contains some long, deep rift valleys, the largest 2200 km long and 5 km deep. One remarkable feature of this map is the apparent absence of evidence for plate tectonics, such as the Venusian equivalent of the mid-ocean ridges on Earth. Some of the slopes look young and smooth and, although the resolution is too poor to be certain, active volcanism is a possibility on Venus. The numbers in the key give the planet radius in kilometres for each colour. The mean radius of the planet is 6051 km. 1 degree equals 106 km.

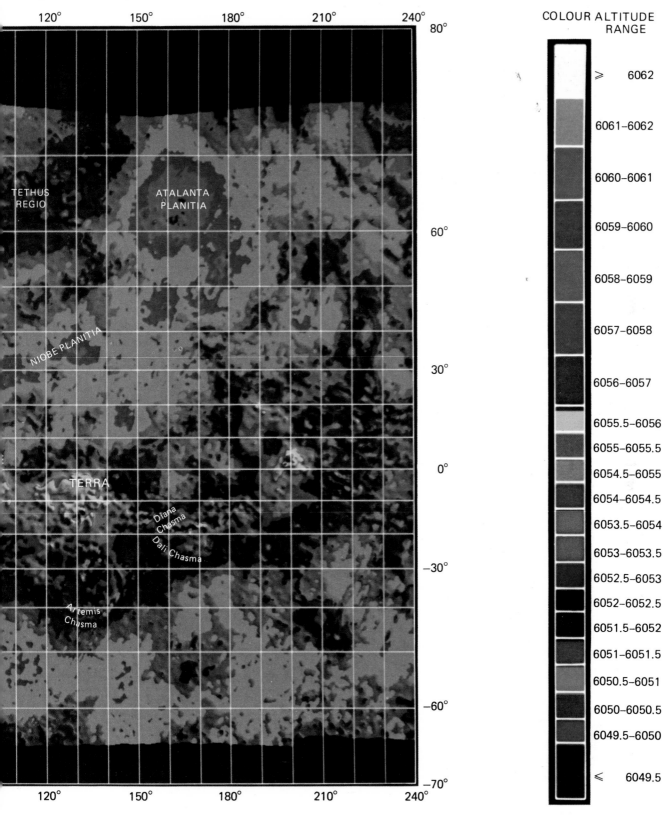

COLOUR ALTITUDE RANGE

	⩾ 6062
	6061–6062
	6060–6061
	6059–6060
	6058–6059
	6057–6058
	6056–6057
	6055.5–6056
	6055–6055.5
	6054.5–6055
	6054–6054.5
	6053.5–6054
	6053–6053.5
	6052.5–6053
	6052–6052.5
	6051.5–6052
	6051–6051.5
	6050.5–6051
	6050–6050.5
	6049.5–6050
	⩽ 6049.5

The cloud markings
which appear when
Venus is viewed
through an ultraviolet
filter are often de-
scribed as resembling
a giant 'Y' laid on its
side. The markings
rotate around the planet
from east to west very
rapidly, completing a
circuit about every 4
days. The search for the
cause of such a rapid
zonal flow on a slowly
rotating planet has
provoked numerous
studies of the general
circulation of Venus's
atmosphere. The
markings also migrate
towards the pole, but
more slowly by a factor
of ten or more. These
pictures were taken at
the Pic du Midi
Observatory in France
in July, 1966 (top) and
from Mariner 10 at a
distance of 3.3 million
kilometres on 10
February, 1974
(bottom).

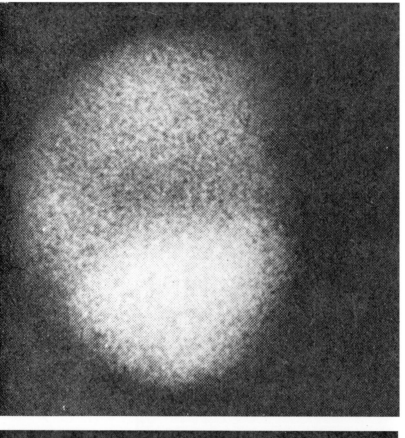

The Russian Venera 9 and 10 orbiters, after they had delivered the landers, also took ultraviolet pictures of the cloud tops. This photograph shows a close-up of a region near the equator as seen from a height of about 1600 km. The markings are now thought to represent variable amounts of sulphur dioxide gas, plus perhaps particles of sulphur itself, swirling around inside the clouds.

Images of the Venus clouds from Mariner 10 (top) and the Earth (bottom) obtained on successive days can be joined together to form time sequences like these. Clearly, waves are propagating around the Venus tropics; the characteristic appearance may be caused by the superposition of bow waves caused by the zonal flow interacting with the turbulent region at the subsolar point. Waves of the same kind of scale are an important part of the Earth's circulation also.

Venus

This drawing, by the Mariner 10 imaging team, shows the names given to the principal semi-permanent features found in the Venus clouds. Examples of each can be seen in the accompanying ultraviolet pictures. The features are manifestations of the circulation of the atmosphere at pressures of around $\frac{1}{10}$ bar, or about 60 km above the surface. Theories of how they originate are discussed in the text.

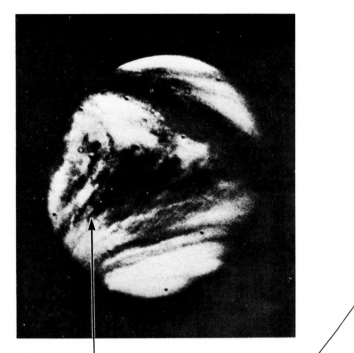

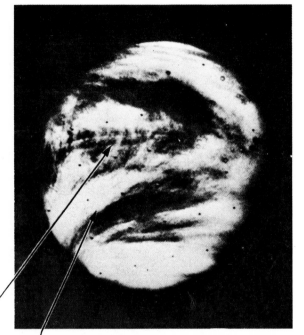

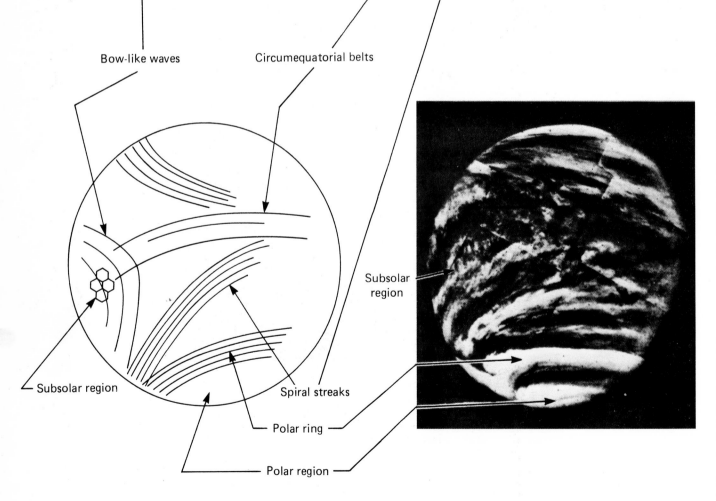

Bow-like waves

Circumequatorial belts

Subsolar region

Subsolar region

Spiral streaks

Polar ring

Polar region

A time-lapsed series of Mariner 10 pictures showing the evolution of one of the features known as a 'circumequatorial belt'. This is the light-coloured, almost horizontal marking near the middle of each picture, which moves towards the south as time advances. Its velocity of propagation works out to be about 20 m/s. Of all the atmospheric features seen on Venus, these are probably the most difficult to explain.

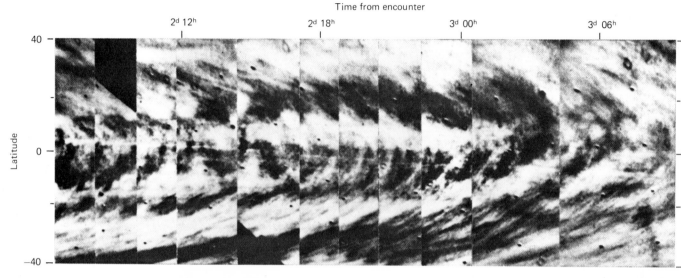

A close-up of 'bow-like waves' seen by Mariner 10. These are probably waves caused by the interaction of the rapid zonal (east to west) flow of the atmosphere with the subsolar disturbance, but a detailed model cannot be constructed until the height distributions and the temperature structure of the markings are better known.

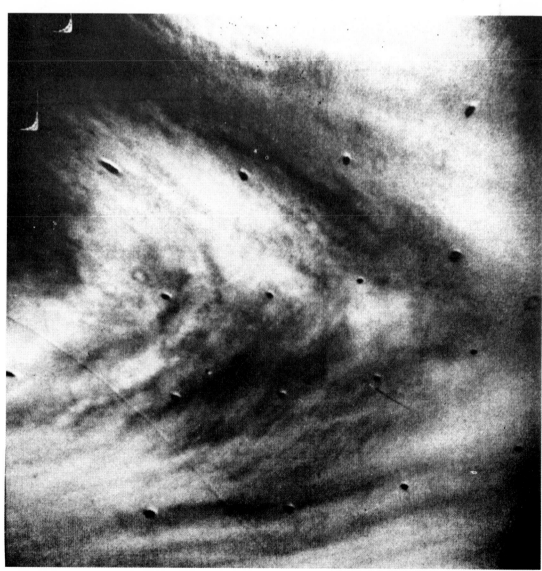

Venus

Two Mariner 10 close-up pictures of the 'subsolar disturbance' on Venus, showing the atmosphere formed into small, vigorous convection cells. The intense sunlight at this location, the local noon point on the equator, penetrates fairly deep into the atmosphere and the resulting heating causes convection which extends up to the level seen here. The process is roughly analogous to that which would be seen if a large, shallow pan of water were heated strongly at one point near the centre. Note that some of the cells are bright-rimmed, and others dark-rimmed. Probably, the difference between the two is that the bright-rimmed cells are 'open' (rising at the centre and spreading out), while the dark-rimmed ones are 'closed' (rising at the edge and descending at the centre). Analogues to these are found in the Earth's atmosphere.

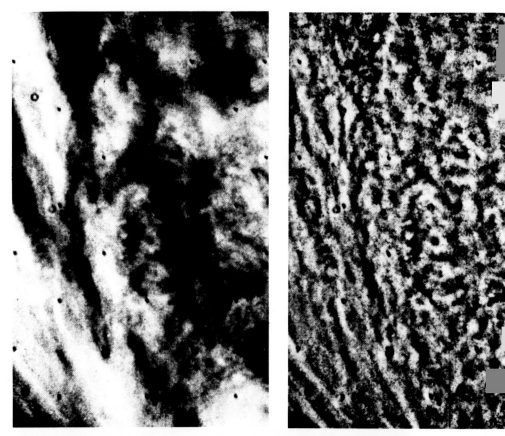

A Pioneer Venus picture, giving a view of the south 'polar ring'. The appearance of the ring is very suggestive of a region of high shear (i.e. gradient in the wind velocity), which implies a discontinuity between two circulation regimes.

66

An idea of the variability of cloud features on Venus
can be gained from this set of four ultraviolet
images, obtained over a period of 38 hours in May,
1980. Note particularly the view of the south polar
vortex at bottom left, and the bright polar rings.
The brightness at high latitudes is due to a greater
number of small particles, which scatter ultraviolet
light more effectively, in the clouds at these
latitudes.

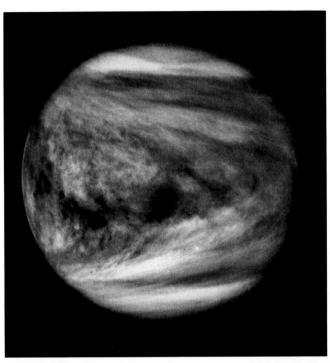

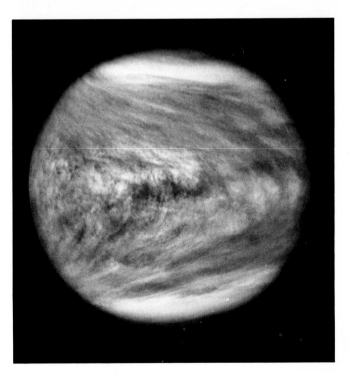

Venus

The phases of Venus as observed by Pioneer between December 1978 and April 1979. Such images are of detailed interest to the scientist, because the dependence on phase of the cloud brightness and polarization yields details of the particle size, concentration and refractive index of the cloud droplets. It was from such data that it was determined that the droplets are spherical (therefore liquid) and have the refractive index of concentrated sulphuric acid.

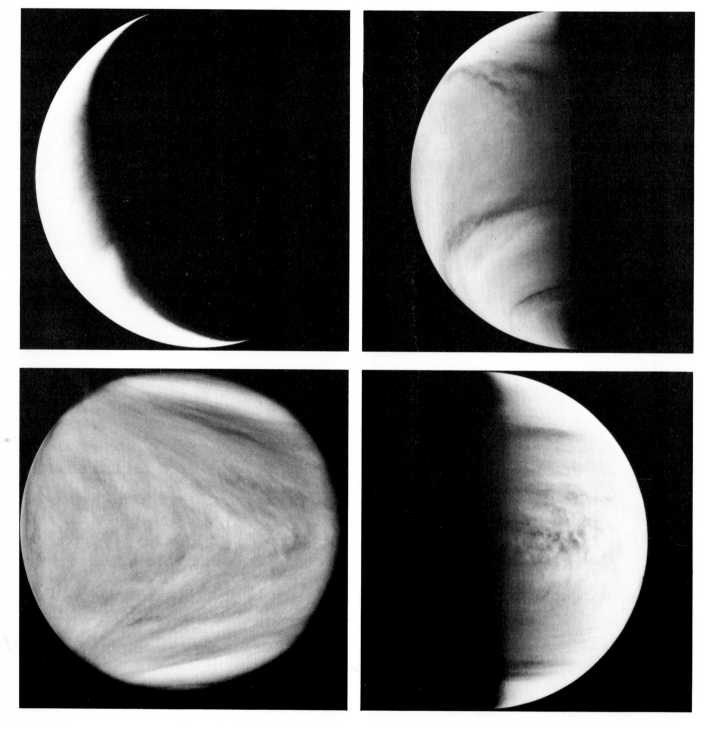

Two polar maps (N pole at the centre, equator around the boundary) of the atmosphere of Venus showing the temperature of the atmosphere at two different heights, (top) about 70 km above the surface (just above the clouds) and (bottom) about 85 km above the surface. At 70 km the most pronounced feature is the circumpolar collar, although the equatorial tides can be seen very clearly also. The polar dipole is smeared out by the averaging process because of its rotation. At the higher level, the dominant feature is the relatively high polar temperatures. These were unexpected, since the solar heating is strongest near the equator. This kind of temperature distribution sets up pressure gradients which oppose the cloud-top winds and so by this level, some 20 km higher, the fast winds are gone. These pictures were built up by averaging together Pioneer Venus infrared data taken over a three-month period. The coldest temperature (blue) is about $-60°C$; the warmest (brown) about $-15°C$.

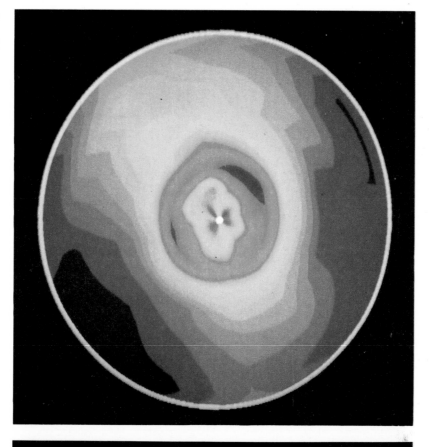

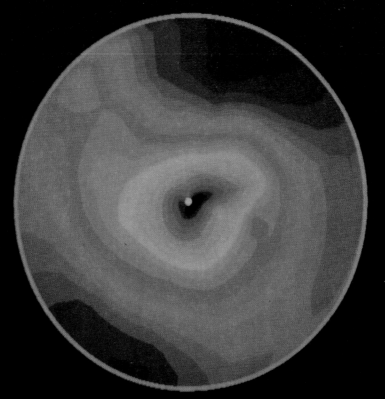

Venus

A picture of Venus as seen from Pioneer in its emitted heat radiation at a wavelength of $11\frac{1}{2}$ micrometres (about 20 times longer than that of visible light). The very bright region in the centre, at the north pole, is due to intense emission from the hot lower atmosphere escaping to space through a region of reduced cloud cover, caused by the atmosphere descending in a 'vortex' pattern analogous to water draining from a bathtub. It is thought that the atmosphere, heated by the Sun, rises at the equator and mid-latitudes and migrates to the pole, where it recirculates itself to the lower levels. Notice the dense cold 'collar' surrounding the pole at high latitudes; this extends above the main cloud deck for about 10 km and is about 1000 km across. It is a planetary wave of some type unknown on the Earth.

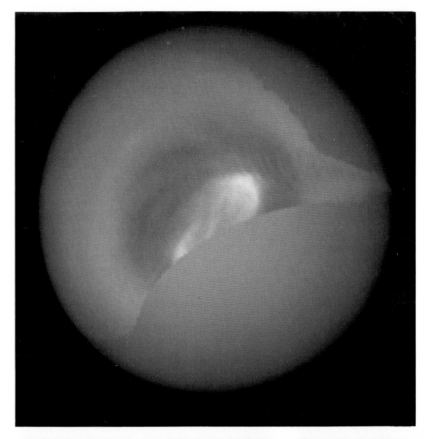

Pioneer infrared pictures looking straight down on the north pole on two different dates (25 December 1978 and 29 January 1979), showing changes in the polar vortex structure. (The pole is at the centre; the edge of the picture is 50°N latitude. The blacked-out region is not the night side – infrared images are indifferent to solar illumination – but is the side of the planet not observed from Pioneer since the orbiter passes to one side of the pole.) The polar cloud structure confirms the expectation of the vortex phenomenon, but shows remarkable and puzzling detail, such as the elongation and doubling of the 'eye' of the vortex and bright S-

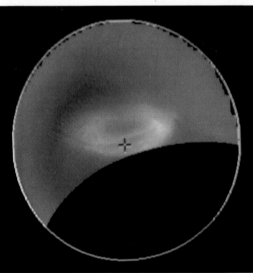

shaped channels in the cloud crossing the pole. Probably, the eye is oval-shaped due to a standing wave pattern with wavenumber-2 around the pole, and it has a 'cap' of high cloud obliterating the centre of the vortex at the pole itself. The dark cloud 'collar' at about 65°N is somehow tied to the position of the Sun in the sky; possibly it is the result of the subsolar disturbance propagating poleward and encountering the edge of the main vortex. It lies further towards the pole than the polar ring feature in the ultraviolet images, but the two are probably related.

An infrared picture of Venus similar to the others but with the contrast 'stretched'. The hot polar dipole and the cold polar collar stand out in this representation, which also brings out two warm regions near the equator (coloured pink). These are the maxima in the equatorial 'thermal tide' on Venus, which unlike the familiar terrestrial equivalent, has two cycles per day. Note also the almost circular isotherm (blue) at mid-latitudes, separating the equatorial and polar dynamical regimes.

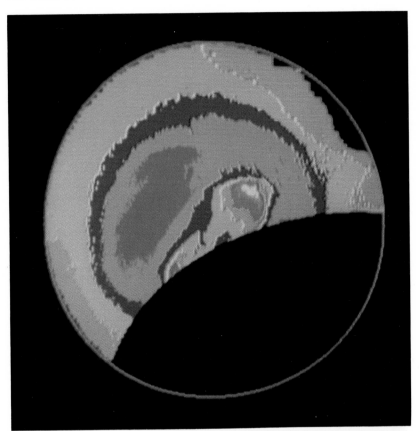

A three-month average of infrared measurements of the north polar region on Venus. In this case, the data from different days is averaged in a coordinate frame which is fixed with respect to the axis of the dipole. In this way the mean structure of the dipole is enhanced and detail in the collar etc. is smeared. The two temperature maxima either side of the pole are clearly seen. These are regions of small or absent cloud cover, which allow infrared emission from the warm atmosphere below to pass through. This kind of clearing is not unexpected – rising air from lower latitudes recycles downwards at the poles and suppresses cloud formation – but the dipolar shape is a fascinating problem.

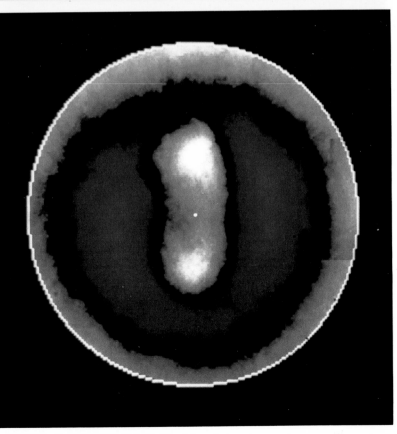

Earth and Moon

Moving outwards from the Sun past Mercury and Venus a traveller would come next to the binary planetary system, Earth/Moon. Nowhere else in the Solar System do we see such a closely matched pair of bodies in association together – although, in fact, the similarities between the two bodies are rather limited on close inspection. It has long been known, from estimations of the masses of the Earth and Moon, that the two are quite different in bulk composition and it has, therefore, been a long-standing puzzle to understand how the two bodies came into conjunction. One of the major scientific goals of the Apollo programme was to shed more light on this mystery and it would be agreeable to report that this goal was achieved. The Moon proved to be more complex and more evolved than some had expected and, although the scientific legacy of Apollo is indeed extraordinary, we are only a little closer to having resolved this problem. After first discussing each body separately the question of how the Earth and Moon entered their celestial partnership will be examined again, and here it is enough to note only that the Earth and Moon do make highly dissimilar companions – the Earth an active planet in every respect and the Moon an airless, geologically senile and biologically sterile world.

Earth Whereas our knowledge of the most intensively explored planets, the Moon and Mars, fills many volumes, our knowledge of the Earth fills libraries. Nevertheless, our understanding of the Earth is largely confined to its present state and its relatively recent history (the past 600 million years or so). Its original state and the first 4 billion years of evolution are known much less well than for any of the other inner planets save Venus (and this will probably change too in the next decade). This discussion of the Earth, therefore, centres on its present state with only a brief speculation about the early evolution. In time our studies of the other planets, where the history of the surface has not been erased in the same way as on Earth, may lead us to a much better comprehension of the complex world on which we live.

The view of Earth from space, available for little more than a decade, shows that our planet is exquisitely beautiful – a jewel-like globe of blue, brown and green with an ever-changing pattern of brilliant white clouds. It is on a global scale that we will discuss the Earth and in the following table are summarized some of its characteristics.

Mass 5.97×10^{27} g	Orbital period 365.26 days
Radius (equatorial) 6378 km	Obliquity $23°27'$
Mean density 5.52 g/cm³	Orbital eccentricity 0.017
Equatorial gravity 978 cm/s²	Mean distance from
Rotational period 23 hr 56 min 4 s	the Sun 149.6×10^6 km

Much of our knowledge of the global properties of our planet has been derived from the analysis of the manner in which the Earth vibrates in response to the release of energy by earthquakes. The behaviour of the different kinds of seismic waves is such that information about the composition and state (solid or liquid, temperature and pressure) can be inferred. It has been determined that the interior is not uniform but that the material is distributed into concentric zones of varying composition and state. Such a situation is predictable from estimates of the Earth's bulk density which is significantly greater than the density of the crustal rocks to which we have access. From this alone it is clear that, somewhere within the Earth, there must be an accumulation of much denser material. Seismic measurements have shown that the Earth has, at its centre, a *core* of very dense material, the outer part of which is in a liquid state. This core is believed to be composed primarily of iron, at temperatures that increase from about 3000°C at the core boundary to about 6000°C at the centre, and has a radius slightly more than half that of the Earth. Self-induced, self-sustaining electrical currents within the core of our rapidly rotating planet are generally supposed to maintain the Earth's magnetic field by way of a dynamo effect.

Most of the rest of the Earth comprises the *mantle,* a shell of hot rock that surrounds the core and extends to within a few tens of kilometres of the solid surface. The mantle is about 2900 km in thickness and contains about two-thirds of the planetary mass. The outermost, solid part of the Earth, that with which we are familiar, is the *crust* and this is little more than a thin, condensed skin on the surface of an incandescent globe. Here, at the crust, the dynamical nature of the hot interior finds expression in numerous ways that will be discussed. The outermost shell, consisting of the Earth's oceans and atmosphere, is also an expression of that dynamism since these volatile substances were evidently produced by outgassing from the interior where they were originally bound.

It is the constant change that we associate with out planet's history that makes the solid body and atmosphere of the Earth so interesting. A good place to start the discussion of this dynamism is by examining further the mantle which provides the driving force for the crustal motions – plate tectonics – that dominate the Earth's geological processes. The composition of the mantle is believed to be primarily iron and magnesium silicates, with relatively small amounts of aluminium, calcium and sodium oxides. Some of the rocks that are found at the Earth's surface have evidently formed at great depth and, in certain locations such as in the igneous intrusions known as kimberlites, it is possible to find rocks having a composition suggestive of mantle origin. Such rocks are largely composed of olivine, a mineral having a chemical composition of $(Fe, Mg)_2SiO_4$.

The mantle is not uniform, for seismic wave velocities show distinct changes at certain depths. It is surmised that the mantle material assumes different crystalline structures depending upon the temperatures and pressures to which it is subjected. Near the mantle/crust boundary, seismic wave velocities in the mantle are relatively low and this is attributed to the presence of trace amounts of water or carbon dioxide. Temperatures inferred for the upper mantle are high enough that partial melting may occur at depths of about 100 km. It is believed that under these circumstances the mantle material is sufficiently plastic to permit convective overturning. The rate of overturning is necessarily extremely slow but the process has critical consequences for the overlying material (this material consists of the crust and the uppermost, rigid mantle which together are termed the *lithosphere*). The plastic properties of rocks at very high temperatures and pressures are not yet well understood, but the idea that the mantle can be deformed and can flow at depths of 100 km or more is generally conceded. This 'soft' part of the upper mantle is termed the *asthenosphere*.

Present thinking about the outer layers of the Earth pictures a rigid lithosphere floating on a hot, plastic asthenosphere which is slowly overturning. This lithosphere does not make up a continuous skin but, rather, is apparently made up of a dozen or so irregularly shaped 'plates', some of which carry continental crust. Slow overturning of the asthenosphere leads to horizontal motion of the lithospheric plates. Exploration of the ocean floors has revealed vast submarine mountain chains in the middle of the oceans. These mid-ocean ridges are apparently the sites of the continuous creation of new lithospheric material by the upwelling of basaltic lava at certain sections of the boundaries between the plates. Volcanism is generally visualized in terms of caldera-topped mountains, but, in fact, this underwater volcanism is by far the greatest source of such activity on the Earth. The steady accumulation of the basaltic rocks along the mid-ocean ridges takes place as neighbouring lithospheric plates are dragged apart, at rates of a few centimetres per year. The basalt, formed by the partial melting of the mantle, rises to the surface along the separating plate boundaries and creates new sea floor.

It is clear that if plates move apart at one boundary they must collide at another. At such colliding plate boundaries one plate may plunge beneath another, transporting lithospheric material back into the mantle. The plunging plate is said to be subducted. The subduction process is a violent one that may lead to the formation of great mountain chains and to volcanism of the most familiar kind. Most major earthquakes occur near plate boundaries and nearly all of the deep earthquakes are associated with subduction. When the collision between two lithospheric plates involves an ocean plate and a continental plate the former is thrust downward because the lighter continental rocks are more buoyant. The descending plate is heated as it descends, primarily by conduction from the surrounding hot mantle and from changes in its crystalline state. It is likely that partial melting of the subducting plate material, including sedimentary rock units that may be dragged along, contributes to the prominent volcanism that is associated with the subduction zones. Extensive chains, or 'arcs', of islands can be built up along the edge of a continent in this manner: the islands of Japan, the Aleutians, the Kuriles, and the Marianas are good examples. The familiar mountain belts at the margins of many continents have a similar origin. The Andes are a prime example, created as a consequence of the collision between the eastward moving Nazca oceanic plate and the South American continental plate. All the associated features can be understood as resulting from this single process: the compressional deformation of continental sedimentary rocks now incorporated into the mountain belt; the creation of the paralleling deep offshore trench; the intrusion of the folded continental rocks by magmas; the elevation of the mountain chain to great heights; the extrusive outpouring of andesitic lavas (named for their widespread occurrence in the Andes); and the continued seismic activity.

In cases where two continental plates collide, both resist being carried down because the plate material is relatively lightweight. The collision process is, accordingly, more complex and typically the crust buckles and thickens to produce great belts of mountains such as the Alps, or great mountainous plateaux like to Himalayas.

On the other planets we have yet to observe any equivalent plate tectonism (although Jupiter's moon Ganymede may have experienced a somewhat analogous crustal process), perhaps because their evolution never progressed to that point, or did not allow the lithosphere to be broken into plates, or perhaps because a stage of plate tectonism has since been obscured by later events. The planet Venus, whose surface is still largely a mystery, is perhaps the best candidate on which to find evidence of plate tectonism because the internal evolution of Venus may have been quite like that of the Earth. Mars does exhibit rifting on a global scale and it does have gigantic volcanoes, but there are no Martian equivalents of folded mountain chains, or ocean ridges or island arcs. The Martian volcanoes are giant shields and it is likely that the upwelling of magma occurred at the sites in question for extended periods leading to the gradual, but in time colossal, accumulation of material. On Earth there are similar, though smaller, volcanoes often far removed from plate boundaries. The underlying activity that leads to this kind of volcanism is also supposed to involve convection within the mantle, but of a different kind from that which apparently leads to plate motion. Isolated, columnar 'plumes' of magma are visualized to underlie these volcanoes, a good example of which is the Hawaiian island chain. It is thought that each of the islands was produced by the same plume as the Pacific plate moved over it: dating studies of these islands and of the Emperor seamounts further to the northwest show a progressive change of age and it is supposed that the plume now underlies the youngest of the islands, Hawaii, which is the site of frequent volcanic eruptions.

The picture that has been described is one of extraordinary geological dynamism: the Earth's surface is constantly changing in major ways as continents move and new mountains form. Equally active are the processes that wear away the mountains and fill in depressions. The continuous cycling of water between oceans and atmosphere provides a highly efficient source of weathering and erosion. Glaciation and winds also add to the erosion. The redeposition of eroded material leads, in time, to new sedimentary rock formations. The twin processes of erosion and deposition are very effective in smoothing out the landscape. This activity, coupled with the continual regeneration of the lithosphere, has largely erased the record of the Earth's earliest history.

The discussion of the global characteristics of the Earth in terms of plate tectonics has the great attraction that it addresses convincingly an array of diverse geological problems and, therefore, this model has come to be generally, if not universally, accepted as a good representation of the Earth's present state. Certainly our world was not always as it is today – the present state is the result of much evolution. Perhaps, like the Moon, the outer few hundred kilometres of the Earth were once molten globally. Such a surface would have cooled relatively rapidly and it is likely that the present high temperature of the Earth's mantle is more the result of the decay of the naturally radioactive elements of uranium, thorium and potassium residing there than it is of the outward flow of heat from the deep interior, stored from the accretionary phase. The early Earth, on the other hand, derived its thermal energy from accretion, perhaps from the decay of short-half-life radioactive elements, and certainly from the exothermic core formation process – itself a product of early intense heating. In this process much of the iron within the Earth was melted and migrated to the core region under the influence of gravity, leading to a final catastrophic displacement of previously existing, less dense, primitive core material and the release of an immense amount of gravitational heat. This early, massive heating did more to shape the internal constitution of the Earth than all subsequent evolutionary events. During this time the Earth underwent major chemical differentiation into a liquid iron and nickel-rich core, a mantle rich in high-pressure silicate minerals, and an outer solid/liquid region enriched in volatile and magmatophile elements. This early differentiation, in fact, was a fundamental event in the Earth's history. It created the conditions that made possible the subsequent evolution of the Earth: the upward migration of lighter elements to form the crust and, eventually, the continental and oceanic rocks; and the upward migration of the very volatile elements that formed the atmosphere and oceans.

The earliest crust of the Earth must have formed and reformed many times. It probably did not consist of the continental and oceanic rocks we know today, but rather was made up of vast accumulations of volcanic rocks. The crust must have undergone the same kind of catastrophic meteorite bombardment that we observe to have occurred on the other inner planets. None of this record remains on Earth, however. The early crust must have formed and sundered many times, due in part to the catastrophic bombardment, but also to recycling driven by convective overturning.

The repeated recycling of the crust and upper mantle continued the differentiation process that led finally to the production of continental nuclei, made stable by the presence of low-volatile-content metamorphic rock bases that inhibited further melting and provided greater strength. The oldest continental rocks yet discovered on Earth, 3.8 billion years old, are from one of these presumed early formed continental nuclei.

Little is known of the grand pattern of plate tectonism throughout Earth history. There are many lines of evidence indicative of the growth of continental plates, involving repeated periods of deformation, metamorphism and igneous activity, but nothing is known of the distribution of the continental masses over the Earth's surface except in relatively recent times. Nor is it likely that the details will ever be known since the creation and destruction of oceanic crust, and the movement of the continental plates, has been so rapid as to preclude ever retrieving such information. The rate of oceanic crust recycling is so rapid in fact that the present ocean floor basalts date back no more than a few hundred million years, about 4% of geologic time. The distribution of continents and ocean basins must have changed continuously as the underlying asthenospheric convection moved the plates, bringing continents together and separating them again. Only the last chapter of this long history is known with some certainty. As recently as two hundred million years ago, after 96% of the Earth's evolution had already taken place, all the present continents were joined together in one supercontinent which has, since that time, broken apart. The separation of those continental fragments continues to this day.

Like the Earth itself, the atmosphere has evolved greatly with time. The present atmosphere is mostly nitrogen together with a significant fraction of oxygen – a gas rarely encountered on other planets where the atmospheric oxygen is generally bound up into carbon dioxide, water vapour and other gaseous oxides. The Earth's free oxygen is a consequence of the origin of life. We do not know what kind of an atmosphere, if any, surrounded the Earth when it had just accreted – one may speculate that it consisted of a mixture of gases captured from the solar nebula and volatiles released from infalling meteoritic materials. Such an early atmosphere would have begun to change immediately as a result of a variety of chemical and physical processes, perhaps including sweeping by an intense early solar wind. We can, however, do little more than speculate about such an evolution at present. The earliest direct information we have is derived from the study of ancient sedimentary rocks: carbonate minerals 3.8 billion years old imply that within a billion years after the Earth's formation the atmosphere included a significant amount of carbon dioxide and hence lacked the highly reducing nature of the solar nebula gases.

It is likely that, over geological time, the gaseous emissions from volcanic activity could have accounted for the Earth's present atmosphere, oceans and the carbon and oxygen locked up in various forms in the crust. The Earth's surface was evidently cool enough early on for the outgassed water vapour to condense so that, in time, the low-lying regions of the crust were covered with liquid water. Whether the continents were in place before the oceans formed is uncertain. It is fortunate that the outgassed water was not much greater than providence ordained since the available area above sea-level could have been very small.

Most of the carbon dioxide that was originally outgassed also no longer resides in the atmosphere: it has been lost by processes that have led to the formation of carbonate deposits (mostly limestone), coal and petroleum. Biological processes have been critical for the limestone formation, which is primarily through the accumulation of the skeletons of tiny sea creatures, and for the deposition of the hydrocarbons, through the burial of vegetable materials. The loss of the water and the carbon dioxide has left nitrogen, a relatively inert gas, as the principal constituent of the air we breath. The oxygen in our atmosphere is thought to be primarily the product of photosynthesis by chlorophyll-containing plants rather than of the inorganic photolysis of atmospheric water vapour. The Earth's present atmosphere is thus seen to be far removed from any that might have evolved in the absence of prolific life. The oxygen is involved in an important cycle: creation by photosynthesis in plants and conversion into carbon dioxide and water in the respiration process of animals. Originally the amount of oxygen in the atmosphere must have been quite small. Oxygen-producing organisms evolved before those which use up oxygen and, on balance, there has been a gradual build up of atmospheric oxygen to a point where about 20% of the air is composed of this vital gas.

The water vapour and carbon dioxide in the Earth's atmosphere are vital for life on this planet both directly, through their involvement in the chemistry of life, and also indirectly in their effect on the Earth's climate. Both gases have molecular characteristics that permit them to absorb the heat radiation (mostly in the infrared) emitted by the warm Earth's surface. This leads to an insulating effect that maintains a significantly higher surface temperature (about 35°C) than would follow from the Sun's heating in the absence of an atmosphere: vital because the temperature increase spans the freezing point of water and provides our world with liquid oceans and a clement climate.

The climate is the average effect of the weather in an atmosphere. The weather is the variable part of the climate produced by the atmospheric circulation, both global and local, and by the variations induced by the orbital motion. The general circulation is produced by differences in heating of the atmosphere from place to place and is a complex topic that can only be touched upon here. On Earth the principal driver of the circulation is the non-uniformity of heating between equator and poles. The simplest kind of circulation that might result from such a variation of heat input might be a single convection cell where warm air rising at the equator would be carried aloft to the poles where it would sink and return to lower latitudes. There is some evidence to suggest that the circulation near the equator has some of these characteristics, though modified by the effects of the Earth's rapid rotation. In the middle latitudes the circulation is much more complex and the transportation of heat polewards appears to be accomplished not by a giant convective cell but rather by large-scale eddies. Evidence for such a motion can be found in global maps of pressure variations and in pictures from space. Such motion can also be simulated in simple rotating 'dish-pan' experiments. It is believed that the Earth's rapid rotation and the relatively intense latitudinal temperature gradient together make the wave regime of circulation the preferred mode for this part of the globe. Meteorological data acquired from Mars indicate that the winter hemisphere circulation there also has a comparable behaviour.

Moon For some time to come, the Moon is likely to remain the only other planetary body that Man has touched and explored himself. Even a generation ago it was almost inconceivable that we might ever journey to the Moon and, as a result, be able to describe confidently what the Moon is made of and what its history has been. In the decade that has followed the Apollo landings an extraordinary depth of knowledge about the Moon has been extracted from the lunar samples (including small samples returned by Soviet spacecraft) and from analyses of the various other kinds of data acquired as a part of the overall Apollo programme. In this quite remarkable effort of scientific analysis new techniques have been invented of the utmost precision and delicacy: the resulting information explosion has helped us to understand much better the Solar System in general, and the early history of the Earth in particular. Surprisingly, however, we have advanced little in our understanding of the most basic question: how and where was the Moon formed. Most of this section, therefore, will address the present state of the Moon and ideas about its evolution. The question of the origin of the Moon will be examined later.

Among the terrestrial planets – Mercury, Venus, Earth, Moon and Mars – the Moon is the smallest and the least dense. It orbits the centre of mass of the Earth–Moon system once every 28 days, and revolves once about its own axis in that time so that we are always presented with the same lunar face. At one time in the past, the Moon was much closer to the Earth and rotated more rapidly than at present; but tides raised in the interior of the Moon have served to transfer rotational angular momentum to orbital angular momentum thereby slowing down the Moon's rotation rate and slowly moving its orbit further from the Earth. Synchronism between the rotational and orbital motions was eventually reached.

Some of the principal global-scale physical characteristics of the Moon are provided in the following table:

Mass 0.073×10^{27} g	Equatorial gravity 162 cm/s
Radius (equatorial) 1738 km	Rotational period 27.3 days
Mean density 3.34 g/cm³	Orbital period 27.3 days

The visible face of the Moon is seen, in reflected sunlight and at full phase, as a bright, white disc with irregular dark patches. It is worth noting that the apparent brightness of the Moon is deceptive – the lunar surface materials are all of low reflectivity and only appear bright because of the contrast of the night sky and because of the sensitivity of the dark-adjusted eye: the dim daytime Moon is more representative of this low-albedo object.

The Moon's geologically active history is now known to have been relatively brief and rather straightforward; the brighter surface regions (termed *highlands*) and the darker regions (called *maria*) hold the clues to the Moon's evolution, once the observer is made aware of what these two classes of surface represent.

Telescopic images, or better yet, those taken from lunar orbit, reveal part of the relationship between the highlands and the mare regions. The highlands, which we now know to represent the original lunar crust, comprise over 80% of the lunar surface (on the far side of the Moon there are few mare regions). They are covered, to the point of saturation, with craters of all sizes and of various morphologies: this cratering produced the basic highland morphology. Some of the craters are immense, of the order of a thousand kilometres across; these are termed 'basins' and they show distinctive forms, typically with an obvious concentric ring-like appearance. Nearly all of these craters are believed to be the result of meteoritic bombardment, much of it taking place during the last stages of accretion. The huge size of the basins indicates that they were formed by the impact of bodies kilometres in radius – so-called 'planetismals'. Ejecta produced by the basin-forming events would have been broadcast globally; around the basin the ejecta materials are thought to have accumulated to an appreciable thickness, covering up the underlying terrain. The morphology of the lunar highlands can be understood almost entirely in terms of the effects of massive meteorite bombardment; these regions are evidently very ancient since the areal density of the craters is so great.

The mare regions, found mostly on the Moon's near side, make a considerable contrast to the rugged highlands. Besides being much darker, the maria are virtually flat and are relatively devoid of craters (indicating that the maria were formed more recently than the highlands). A long-held suspicion about the nature of the maria has been confirmed by lunar sample analysis, namely, that these are regions of lava flooding.

Samples of both highland and mare materials now reside in numerous laboratories and approximately 2000 transactions take place each year at the curatorial facility at Johnson Space Center in Houston where the U.S. lunar samples are housed when they are not on loan to qualified investigators across the world. The Apollo landing sites, from which all the U.S. samples were returned, were selected only after many factors, especially scientific value and landing safety, had been taken into account. The maria are the least rugged and hazardous regions on the Moon, and the first two landings were in such areas: Apollo 11 touched down in Mare Tranquillitatis and Apollo 12 in Oceanus Procellarum. The third landing, Apollo 14, was on the terrain known as the Frau Mauro formation, thought to be material ejected from the great Imbrium Basin. The last three Apollos sampled more diverse terrains. Apollo 15 landed near Rima Hadley, a sinuous depression in a mare plain at the eastern edge of the Imbrium Basin: the site was just inside the Apennine Mountains, which are thought to be fault blocks uplifted at the time of the Imbrium event. The Apollo 16 site was the only one in the lunar highlands – in a region named Descartes, about 400 kilometres southwest of the Apollo 11 site, while Apollo 17 landed in the Taurus–Littrow region on the rim of the lava-filled basin, Mare Serenetatis. The astronauts brought back hundreds of carefully selected and documented samples from the regions near where the lunar modules touched down. In addition, as has already been noted, much smaller quantities of lunar materials, extracted from the surface by a coring device, have been returned by the highly successful Soviet Luna spacecraft: these have sampled a variety of terrains, including the highlands.

The returned samples provide a representative sample of the variety of materials thought to be found across the Moon. The precious rocks are to be found in laboratories in many countries and have been subjected to a great range of physical and chemical analyses that have disclosed the individual elemental, mineralogical and physical characteristics of each, including the ages of the samples. The latter measurement, made by determining the abundances of particular radioactive mother and daughter elements, is a key one for the elucidation of the sequence of events to which the Moon has been subject. From the results of these analyses, from the study of a variety of data acquired from spacecraft in orbit about the Moon, and from the information returned from geophysical stations erected on the Moon by the Apollo astronauts, we now have a satisfying picture of how, in general, the Moon has evolved.

Chemical analysis of samples of igneous rocks returned from the Moon indicates that there are three principal kinds: ferroan anorthosites, norites (sometimes called simply Mg-rich rocks) and basalts. Whereas the elemental composition of a rock tells little about the genesis of the rock, the mineral composition is highly diagnostic and it can therefore be confidently asserted that the noritic and basaltic materials have a quite different origin from the anorthosites: the former are the products of the *partial melting* of mantle materials, while the latter is the result of *crystal fractionation*. Partial melting is the process by which most terrestrial volcanism occurs: mantle materials subjected to heating at depth undergo the melting of those materials that have a low melting point, and the molten rock migrates to the surface where it cools and solidifies. Crystal fractionation, on the other hand, occurs when total melting of the material in question takes place followed by subsequent cooling; the discovery that the lunar highlands, which cover most of the Moon's surface, are derived to a great extent from such a process, has profound implications that will be outlined.

The anorthositic highland rocks do not have a low-melting-point composition and were apparently produced from an entirely molten parent material. Anorthosites contain a relatively large abundance of lightweight metals, aluminium and calcium, and appear to have been formed by the flotation of low-density crystals as cooling occurred (or, equivalently, the sinking of more-dense still-molten material). The highlands extend over most of the Moon and presumably underly the younger maria. Seismic data from the geophysical stations left at the Apollo landing sites indicate that the Moon's crust – the highlands – is 50 to 100 kilometres in thickness. The community of lunar scientists has been forced to conclude that at one time the entire surface of the Moon was molten, to depths of several hundred kilometres, and that from this magma ocean the anorthositic crust formed. Age determinations indicate that the anorthosites are older than the norites and appreciably older than the basalts which were, apparently, formed later as discrete igneous events.

The concept of global surface melting is arresting; the same process may have been common to all the terrestrial planets, including the Earth. The source of heat responsible for the wholesale surface melting is, as yet, uncertain. The saturation cratering of the highlands indicates that the meteoritic bombardment that accompanied the late stages of lunar formation was intense; the heat produced by the impact events might have been sufficient if the time-scale of accretion was short enough so that the heat produced did not have enough time to be radiated away. It is estimated that if the time-scale were as short as a thousand years then gravitational (impact) heating could have been a sufficient source to have caused the global melting. Other sources of heating have been suggested including electromagnetic heating by induction from a rapidly varying and intense solar magnetic field. Short-lived, and now extinct, radioactive elements may have had a role as well.

The norite samples returned from the Moon appear to originate in isolated areas within the highlands. Ages derived for these samples indicate that they are a few hundred million years younger than the highland anorthosites (these are dated at about 4.2 to 4.5 billion years). The molten material from which the norites formed was apparently created at depth and migrated upward. The norites are characteristically enriched in radioactive elements (uranium, thorium and potassium-40) and other rare elements. On two of the Apollo missions (15 and 16) measurements were made of gamma-ray emissions from the surface, and the results showed that certain localities contained enhanced levels of radioactivity. These are thought to be regions where the surface materials are enriched with norites. A region between Mare Imbrium and Oceanus Procellarum is one such region. Although these regions are quite localized it was not mere chance that allowed the Apollo astronauts to collect samples of norite material – impact events scatter surface material far and wide, allowing representative samples of lunar rocks to be gathered almost anywhere on the Moon.

The lunar basalts have the youngest ages, generally in the range 3 to 4 billion years, although photogeological studies using crater densities suggest that some maria are significantly younger. As is the case with terrestrial basalts, these lunar samples appear to originate as a result of partial melting in the upper mantle. The maria fill the largest depressions on the Moon – the basins. Repeated eruptions apparently filled the basins to depths of several kilometres. Some of the most recent flows can be observed and mapped in photographs of the *maria* taken from orbit.

Since the formation of the maria there has been little geological evolution on the Moon other than the formation of more impact craters, at an ever-diminishing rate. The lunar history can, therefore, be summarized as consisting of five principal episodes: (1) accretion and subsequent global melting; (2) crustal separation and concurrent massive meteorite bombardment; (3) partial melting at depth and the emplacement of the norites; (4) diminishing bombardment, further melting at depth and the subsequent emplacement of the *mare* basalts; and (5) cessation of volcanism and gradual internal cooling.

Evidence concerning the present state of the interior of the Moon has been obtained from the analysis of seismic data gathered by the geophysical stations left at the Apollo landing sites. These data indicate that the Moon's interior is solid to a depth of about 1000 kilometres, with, perhaps, a still-molten centre. The seismicity of the Moon is extremely low but, because the windless surface of the Moon is an ideal location for a seismometer, it was possible to emplace instruments of the greatest sensitivity and record large numbers of very low-level quakes. Most of these originate at great depth – apparently, near the boundary of the lithosphere (solid zone) and the less rigid central part. The quakes show a 28 day periodicity, a result of the tidal interaction with the Earth.

The figure of the Moon is highly asymmetric – a fact long known – with the centre of the Moon's figure displaced about two kilometres from the centre of mass. The simplest interpretation of this observation is that the (lightweight) crust on the Earth-facing side of the Moon is thinner than that on the far side (the centre of mass is displaced towards the Earth and to the east). This asymmetry may be the result of a chemical inhomogeneity at the time of accretion: this, ultimately, may have led to the formation of maria on the near side of the Moon and their virtual absence on the far side.

Just as the interior of the Moon is that of a 'dead' planet so is the surface geologically and biologically dead. There are no organic materials to be found in any of the returned samples – a finding that is not too surprising on an airless, waterless body that is subject to unabated solar irradiation. The almost total absence of any chemically bound water in the returned samples was, perhaps, more surprising. The Moon is highly deficient in all volatiles in comparison to the Earth, including all those chemical elements that are more volatile than iron. Iron itself is also highly deficient, an important fact that had been inferred long before the space age, on the basis of lunar mass estimates. The volatile deficiency is accompanied by a considerable enrichment of refractory materials (those with high melting points) – including calcium, aluminium and titanium oxides, and minor elements such as barium, strontium, uranium and thorium. The composition of the Moon is, therefore, quite different from that of the Earth. We still do not have a clear understanding of how the Earth–Moon system came to be composed of such a dissimilar pair of bodies.

Origin of the Moon

The problem of the origin of the Moon had been raised, and a variety of hypotheses proposed, long before the present age of spaceflight and lunar landings. An obvious possibility is that the Earth and the Moon were formed as twins. If so, then how does one account for the large difference in density between the two bodies? And why is the orbit of the Moon about the Earth inclined at an angle to both the plane of the Earth's orbit and the equator? Perhaps the Moon formed in quite a different part of the Solar System and was subsequently captured. In such a case, both the means by which the encounter between Earth and Moon took place and the mechanism of capture are unclear: the body to be captured must lose a significant amount of energy if it is to be slowed enough to go into orbit about the second body – unless the satellite-to-be is hot and easily deformed, the loss of sufficient energy cannot occur. One theory of the Moon's origin that neatly accounts for the density difference holds that the Moon was spun off from the Earth after the formation of the Earth's core. In this model the Moon is formed from iron-depleted mantle material. This *fission* model is, however, not in accord with the observed angular momentum of the Earth–Moon system: angular momentum is a *conserved* quantity, and the present angular momentum of the system is a factor of 2 too small to have allowed fission.

The results of lunar exploration to date have not settled the question of how the Moon was formed: none of the three hypotheses mentioned has been confirmed nor have sufficient constraints been forthcoming to entirely discount any one. If the Moon had been found to have a geochemical character fundamentally different from that of the Earth (e.g. in terms of the relative abundances of the isotopes of, say, oxygen) it would have been necessary to conclude that the Moon had originated far from the Earth: the capture hypothesis would have been indicated. In fact, the analysis of lunar samples shows that, although the Moon is deficient in iron and related elements and is enriched in refractory elements like titanium, the isotopic patterns are the same. This does not necessarily imply that the Earth and Moon were formed in each other's neighbourhood but simply fails to confirm distant formation.

The Apollo results might have supported the fission hypothesis if the composition of the Moon had been found to be similar to that of the Earth's mantle. However, the returned samples of mare basalts, which were formed by melting at depth and, therefore, provide evidence of the lunar interior composition, show that the basalts were derived from materials rather different from those of the Earth's mantle. So, the fission concept is not supported directly.

The possibility that Earth and Moon might have formed as twin planets is not a concept for which the lunar sample studies could provide a positive test. So, with all of the basic problems associated with each hypothesis intact, and with no compelling clue from the lunar sample analyses, the question of the Moon's origin remains unresolved. A more complex mechanism than any one of the three discussed is most likely called for.

One idea that has been proposed calls for the proto-Earth to capture a large number of relatively small objects and for the Moon to accrete subsequently from this debris. The assumption is made that the captured material must have already undergone differentiation into metallic iron and silicates: the more easily deformed silicate material would be more easily captured than the iron-rich fragments so that the necessary fractionation could be achieved. Other ingenious ideas have been proposed but none has the virtue of simplicity and susceptibility to ready test.

A widely accepted theory of the Moon's origin may continue to elude us. This should not, however, cause one to underestimate the extraordinary achievements of lunar exploration to date: the evolutionary path of the Moon is now known better than the Earth's!

Innumerable spectacular pictures of the Moon and Planets have been returned over the last two decades but none is more beautiful than the view of our own planet Earth when seen as a whole against the blackness of space. Everchanging cloud patterns, both complex and delicate, superimposed on the blue oceans and brown and green continents make views of the Earth, like this one seen by the returning Apollo 11 astronauts, endlessly fascinating. This photograph was taken over the Pacific Ocean and emphasizes the predominance of water at the Earth's surface. The western United States can just be seen at top right. The cloud pattern is typical: a broad band of clouds at the equator, relatively clear air in the tropics and more complex structure in middle latitudes.

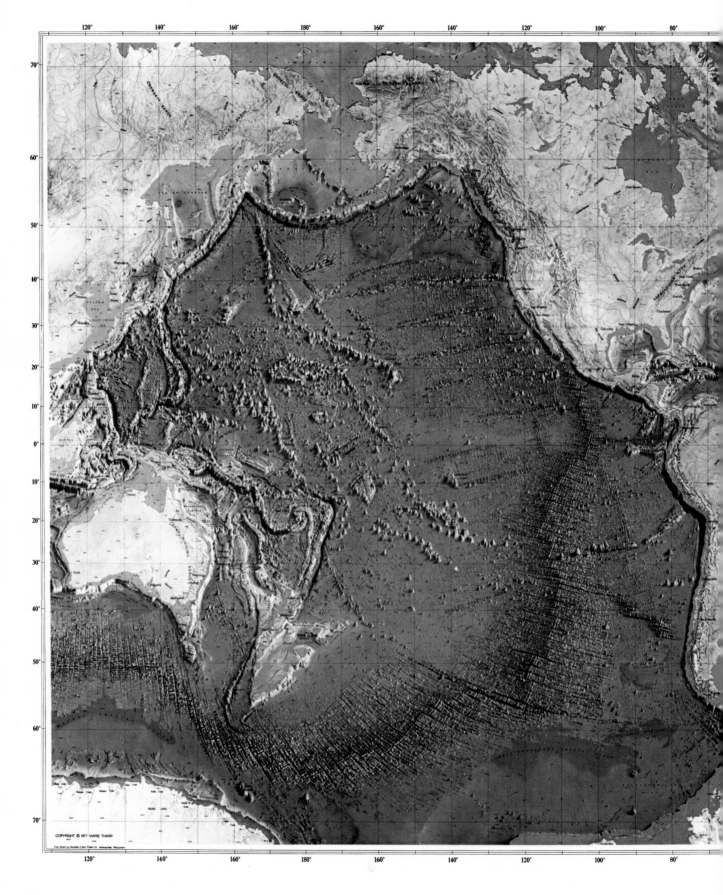

COPYRIGHT © 1977 MARIE THARP

Film Work by Mueller Color Plate Co. Milwaukee, Wisconsin

This map shows clearly the way the surface of the
Earth is made up of continental plates. 1 degree
equals 111 km. It begins to look as though the
Earth may be unique in the Solar System in this
respect. A possible exception is the Jovian satellite
Ganymede.

WORLD OCEAN FLOOR
BY BRUCE C. HEEZEN AND MARIE THARP

Based on Research and Exploration Initiated and Supported by the

UNITED STATES NAVY OFFICE OF NAVAL RESEARCH

1977

A striking observation of a large stationary thunderstorm cell, located over the jungles of South America, was made by the Apollo 9 astronauts on one of the proving flights that preceded the lunar missions. The cell is about 100 km across. Tropical rainforests are exceptionally moist and thunderstorms are frequent – moisture is lifted by strong updraughts in the centre of the storm and the condensing water spreads outwards at altitude, in this case in an especially symmetric pattern.

Hurricane Gladys off the Florida coast. Hurricanes are one of the most destructive of natural phenomena and the advent of weather monitoring from space has greatly alleviated the loss of life caused by these violent tropical storms. This particular picture is not a routine product of a meteorological satellite, however, but was acquired by an Apollo 7 astronaut in October 1968 using a hand-held camera. Hurricanes (called *typhoons* in the Pacific) are produced in circumstances where strong rising currents of air in a central cell suck up moisture, usually from warm ocean water, that condenses at altitude. This condensation produces substantial latent heat release to provide energy to drive the storm. Cyclonic winds develop around the low pressure core: reaching speeds of over 150 km/hour these powerful winds contribute greatly to the hurricane's destructiveness.

Earth and Moon

Part of Africa's longest river, the Nile, is seen in this oblique view taken from the Apollo 9 spacecraft when it was passing over the Sudan. Most of the segment of the Nile seen in this view, looking to the northeast over Upper Egypt towards the Red Sea, is in fact a manmade lake, Lake Nasser, created by the building of the Aswan High Dam. The territory on either side of the Nile is among the most hostile in the world with the Sahara to the west and the Nubian Desert to the east.

A small part of the Antarctic continent
(Victoria Land) is shown in this LANDSAT frame,
measuring 180 km on a side. The small arrow in this
and later Earth pictures points north. Thick,
permanent water ice glaciers have built up on the
piece of continental plate which drifted to the
vicinity of the south pole about 200 million years
ago. Antarctica now contains, in frozen form, most
of the Earth's fresh water supply. The situation at
the opposite pole is quite different: the Arctic ice
cap does not sit on continental crust but, rather,
floats on the ocean. In recent years Antarctica has
proved to be a remarkable source of large numbers
of meteorites. The meteorites have fallen to Earth
over the last several million years and have been
concentrated in certain regions by natural glacial
action. These extraterrestrial rocks are more readily
detected against the pristine whiteness of the ice
than in other terrestrial locations. In time the
analysis of these meteorites is expected to cast more
light on the processes that led to the formation of the
Solar System.

Earth and Moon

Manicouagan Lake, Quebec, Canada. This circular feature located in a pre-Cambrian shield area in eastern Canada is a possible example of an ancient terrestrial impact crater, comparable in size (65 km across) to some of the large craters on the Moon and other planets. A second possibility is that it is an ancient caldera – if so it is of unusually large size. The picture was acquired by LANDSAT.

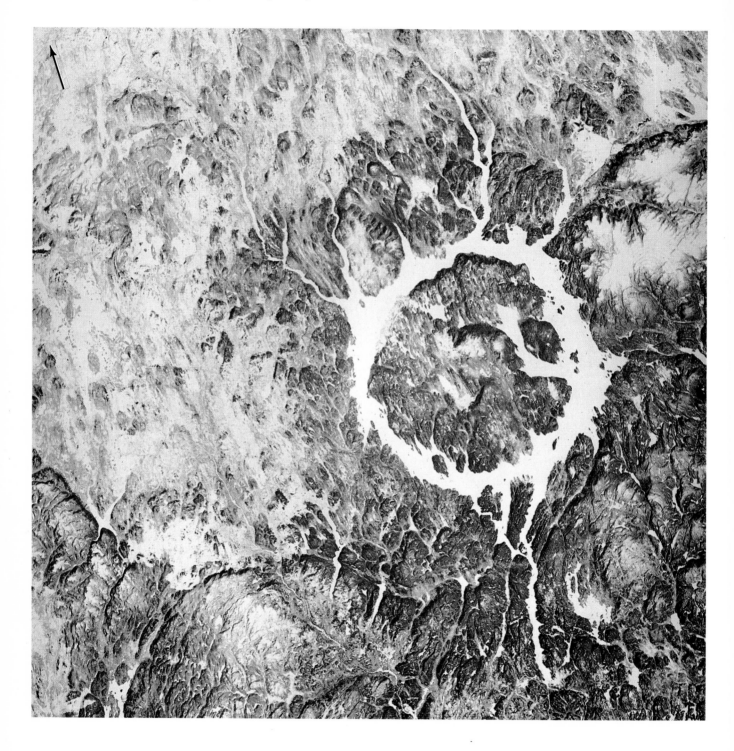

The results of the collision of one crustal plate, the Indian subcontinent, with another, the Asian Plate, are seen in this false-colour LANDSAT frame. The picture shows the contact between the high plains of the Ganges River and the Himalaya mountains of Nepal. In a collision between an oceanic plate and a less-dense continental plate, the oceanic plate is subducted in a way that does not lead to the creation of a massive mountain belt. Here, two continental plates are in collision and the result is the upthrusting of a huge mass of faulted rock, now weathered to form the Earth's greatest mountain region. Mount Everest is in the top right hand corner of the picture.

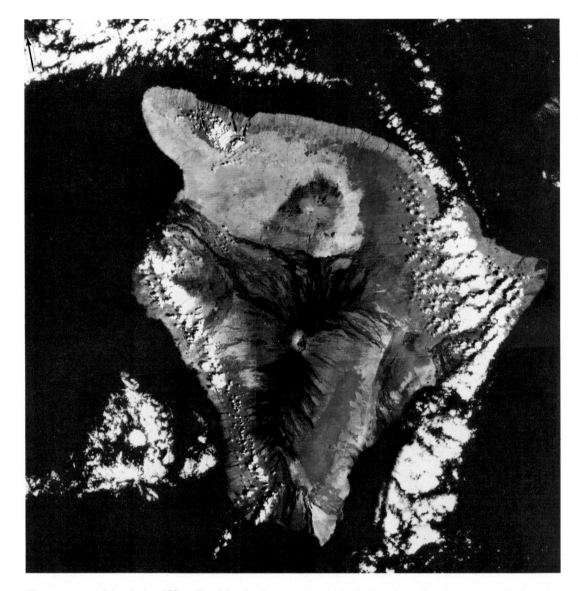

The youngest of the chain of Hawaiian islands, the Island of Hawaii was formed, like the others, by the eruption, from the ocean floor, of basaltic lavas which accumulated to form a giant shield volcano. As such, this shield probably represents the best terrestrial analogue to the colossal volcanic shields on Mars. The top of the highest caldera on the island is 9 km above the ocean floor and, at its base on the ocean floor, is 120 km wide. The dimensions of the Martian volcanoes are 3–5 times as great. The Hawaiian islands are located in the *middle* of the Earth's Pacific Plate. The source of magma that created the islands, therefore, is not the upwelling of lava that occurs at the mid-oceanic ridges at the plate boundaries where the plates are pushed apart. Rather, the source appears to be a hot region in the mantle underlying the crust which creates a magma 'plume'. As the crustal plate slowly moves over the stationary plume a chain of volcanoes is formed. This picture of Hawaii's 'Big Island' comprises two LANDSAT images reconstructed in false colour where a green image has been projected through a blue filter, a red image has been projected through a green filter, and an infrared image has been projected through a red filter. In false colour, vegetation appears in shades of red, rocks in shades of blue, yellow and brown, and water appears as dark blue or black.

Portions of Egypt, Israel, Jordan and Saudi Arabia, as well as the Sinai Peninsula, are visible in this LANDSAT mosaic. The region is one of unusual geologic interest because the Red Sea is thought to be a widening rift that is separating the Arabian Plate from the African Plate. The rupture is believed to have occurred about 35 million years ago.

Earth and Moon

The Great Glen in Scotland is one of the Earth's more readily visible faults: the fault is the valley running near-vertically to the left hand side of this LANDSAT frame. It cuts through the snow-covered highlands and stretches from the North Sea to the Atlantic. More than 100 km of lateral motion along the fault has taken place, mostly over 200 million years ago. Small earthquakes are still recorded here, in a part of the world not well known for tectonic activity.

The Grand Canyon in Arizona is justly famous for its size and grandeur. Although there is a natural tendency for a comparison to be made between this canyon system and the Valles Marineris on Mars, they are in fact entirely different in scale and in origin. Whereas the Martian canyon system was created originally by planetary scale tectonism, the Arizona canyon was formed by the cutting of the powerful Colorado River into the sedimentary deposits of the plateau. The Grand Canyon is about 1.5 km deep and about 20 km from rim to rim. This LANDSAT picture, covering most of the system of canyons that make up the Grand Canyon, is about 175 km on a side. Comparable figures for the Valles Marineris are: 6 km deep, 200–800 km wide, and 5000 km long. The white areas in this picture are snow-covered regions of high elevation. The one at the bottom right is a volcano field – the 3.7 km (12 000 ft) high San Francisco Peaks.

Earth and Moon

The most notable observations of active volcanism from spacecraft are undoubtedly contained in the Voyager images of Jupiter's moon Io. Pictures of active volcanoes on Earth have also been taken from space: this LANDSAT false-colour image records the eruption of Tiatia in July 1973. The volcano is located on the southernmost of the Kuril Islands that run between Japan and eastern Siberia – an excellent example of an island arc created by plate tectonics. The prevailing winds are carrying a giant plume of ash and volcanic gases

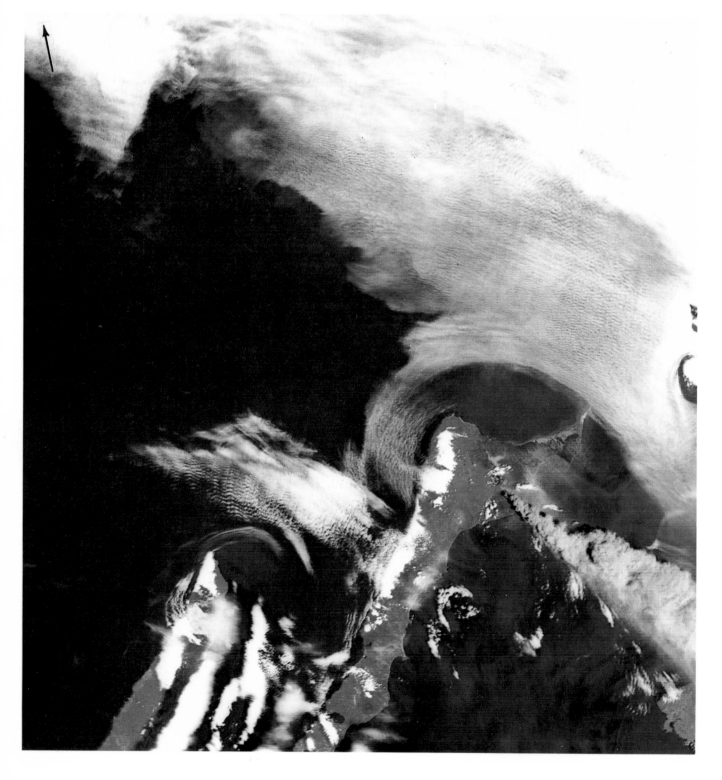

This single LANDSAT frame (180 km on a side) was taken a few hundred kilometres to the east of the area shown in the earlier photograph of the Sinai Peninsula and shows part of the southern Arabian Peninsula. It provides an excellent example of the drainage pattern of a terrestrial desert region for comparison with the hypothesized Martian case. Also of interest are the long parallel seif dunes in the upper left of the frame: these are located in the Rub' at Khali Desert and are formed in response to predominantly northeasterly winds. At the top centre of the frame these dunes merge into smaller individual dunes that resemble more those found on Mars.

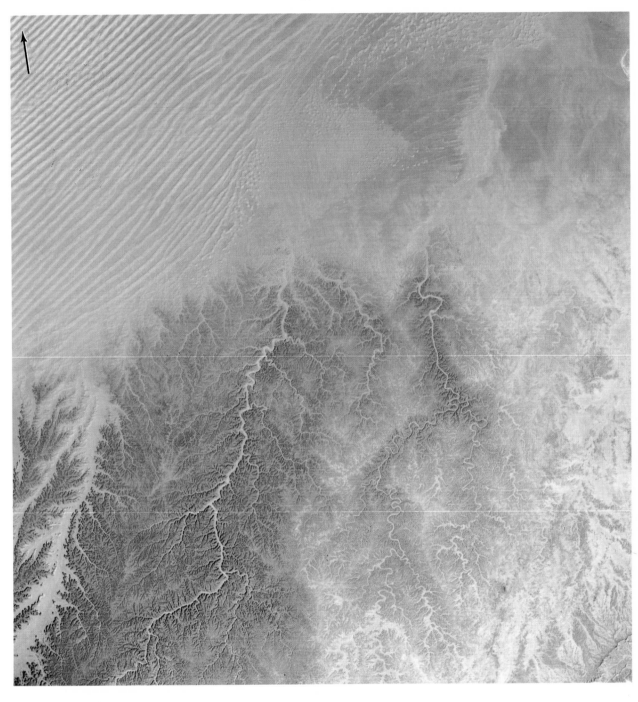

Earth and Moon

The two global views of the Moon shown here and over the page are the products of airbrush artists working from photographs: they show the major physiographic features that will be discussed in more detail on the following pages. The near side of the Moon shown here (the only side visible from Earth), with its prominent smooth mare regions differs markedly from the far side (shown over the page) which is more or less uniformly saturated with impact craters and is almost entirely lacking in maria. The reason for this dichotomy is not yet well understood but is evidently related to differences in crustal thicknesses on the near and far sides, differences that arose early in the Moon's history. The far side of the Moon was first photographed by the Soviet Lunar 3 spacecraft in 1959 but

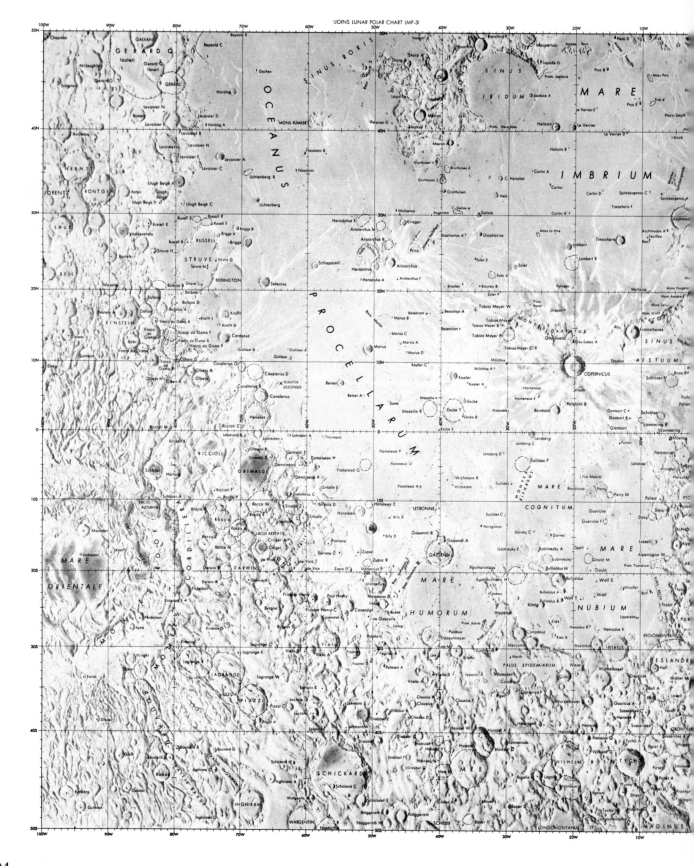

high-quality imaging of
the entire far side was
not acquired until the
Lunar Orbiter and
Apollo flights at the end
of the 1960s. (1 degree
equals 30 km.)

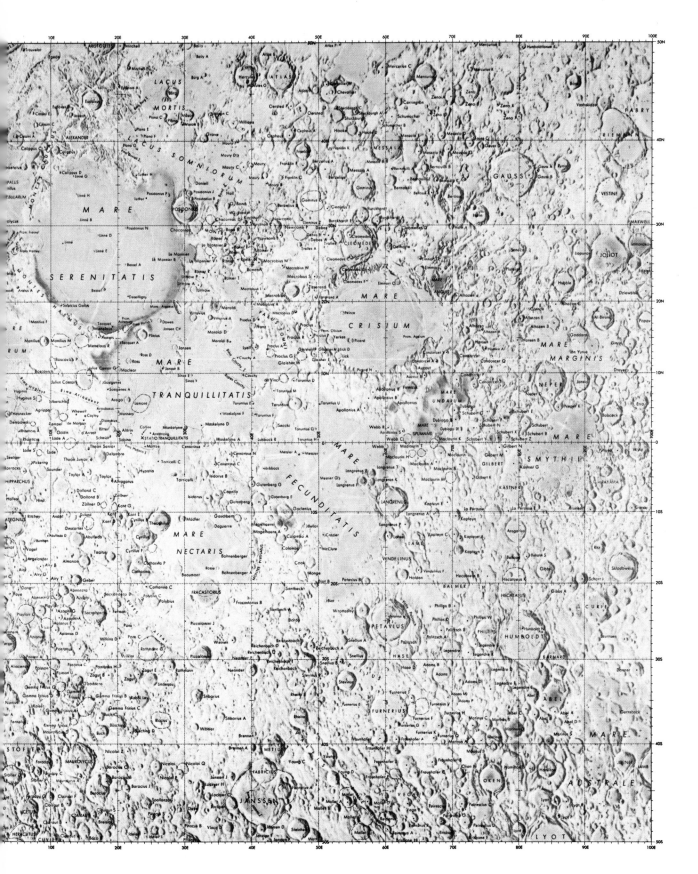

105

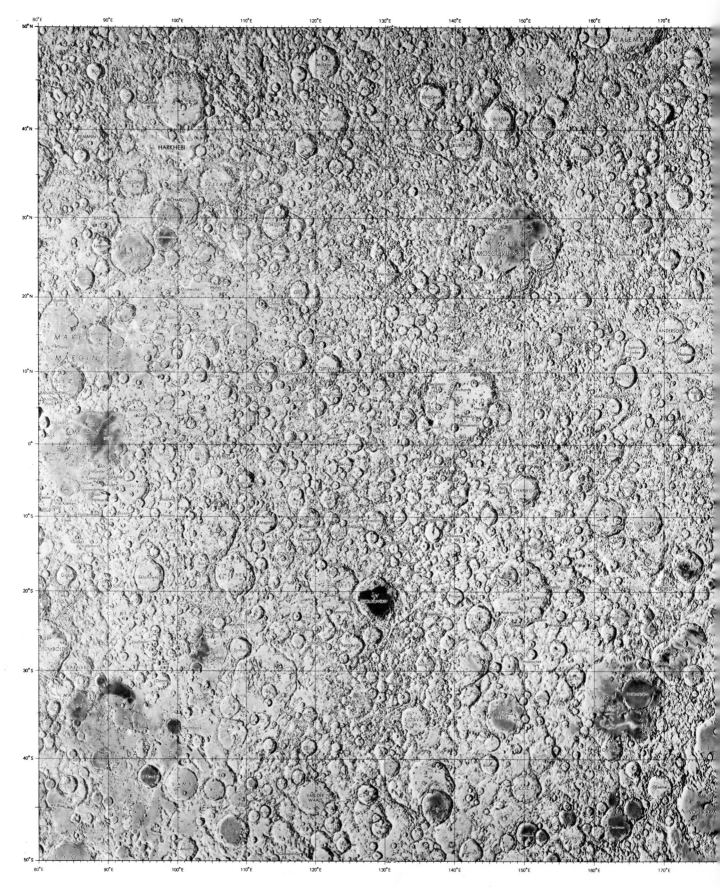

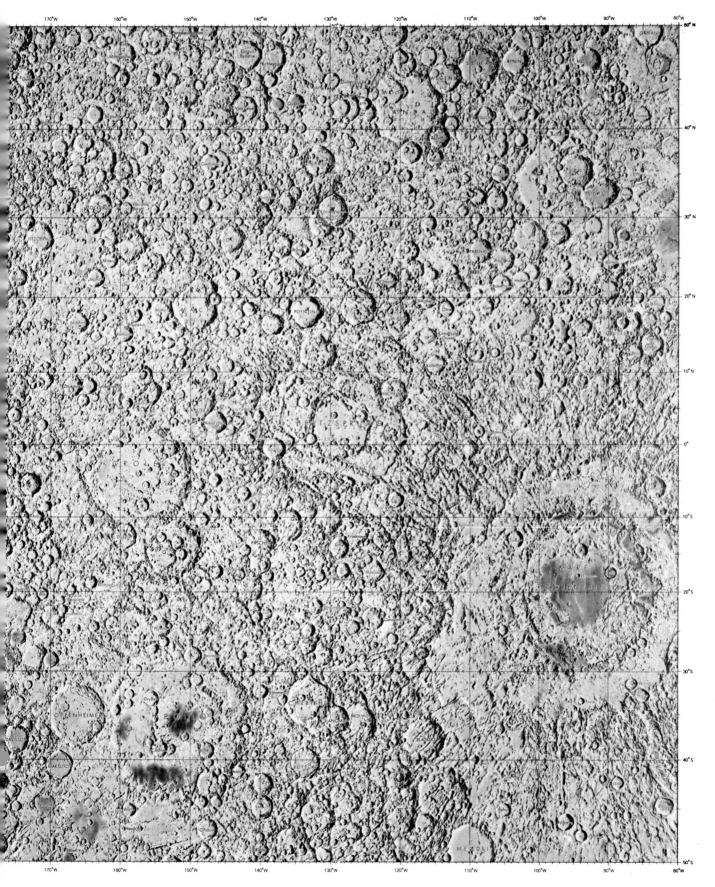

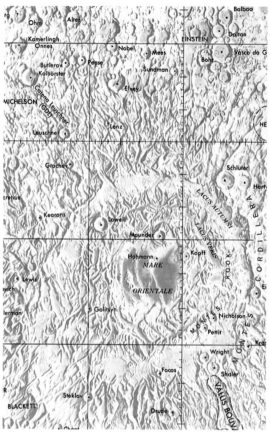

The interesting question has been raised by an astronomer of how the course of human affairs might have been altered if the great unblinking 'bull's eye' that the Orientale Basin resembles had been formed in the middle of the lunar near side rather than in its actual position on the western limb! Orientale was among the last of the cataclysmic basin-forming events to occur on the Moon – somewhat less than 4 billion years ago. Orientale exhibits the classic multi-ring structure of a large basin and has been preserved with relatively little modification. Whereas the other large basins have been flooded with lavas, Orientale has only modest *mare* inundation in its centre. Enormous amounts of debris were thrown out of the basin: the thickness of the ejecta blanket has been estimated to be about 4 km at a distance of 300 km from Orientale's centre. The outermost scarp of the basin, almost 900 km in diameter, rises about 6 km above the surrounding plains. This mountain ring is named the Cordillera Mountains. Lunar Orbiter 4 took this remarkable composite picture in 1967. The horizontal lines mark the joins between the many adjacent scans which have been mosaiced to form the view.

This picture was acquired by a film camera in the service module of the Apollo 16 spacecraft and covers a region on the near side of the Moon near the equator. North is to the top left. The region in question includes part of the *mare* Oceanus Procellarum and, on the right-hand side, Mare Cognitum. Bisecting the area is a rugged highland region, Montes Riphaeus, which has survived the lava flooding which produced this vast volcanic plain. The landing sites of three U.S. space-craft – Surveyor 3, Apollo 12, and Apollo 14 – are contained within this view. Samples collected at the Apollo 14 site include material that is believed to have been ejected by the impact that created the Imbrium Basin nearly 1200 km to the north.

This northward-looking oblique view acquired from the Apollo 17 spacecraft shows the western side of Mare Serenetatis which is bounded by the Apennine Mountains. Within the lava-flooded basin can be seen numerous elevated elongate features which are termed *mare ridges*. The origin of these ridges is not well understood but they are evidently the result of crustal compression, perhaps the surface expression of thrust faults. The ridges are apparently contemporaneous with the episodes that emplaced the *mare* lavas. Also prominent in the lower half of the picture are elongate depressions within the *mare*. These fault valleys, called *graben* are the result of crustal entension: long cracks open up in the crust and a wedge-shaped block collapses into the gap. Typically lunar graben are 1–2 km wide and tens to hundreds of kilometres in length.

There are few *maria* on the far side of the Moon. The volcanic flooding of dark *mare* material (basaltic lava) on the floor of the large (190 km diameter) crater Tsiolkovsky, therefore, makes this one of the most prominent features on the lunar far side. In many respects Tsiolkovsky typifies a crater in its size range with its terraced walls and central peak. It is unclear why this particular crater was the site of the volcanism that led to the inundation. In these two views, which were acquired from the Apollo 15 spacecraft, north is to the right. The upper photograph shows additional detail of the northwestern rim of the crater which cannot be seen in the oblique, overall view below. A broad apron of striated material, extending for over 50 km and demarked by a distinct lobate front, appears to be an enormous landslide. The origin of the slide is evidently a scarp on the outer rim of Tsiolkovsky. The vertical relief from the scarp to the lobate front is about 3 km. This lunar case should be compared with the landslides that have been observed on Mars.

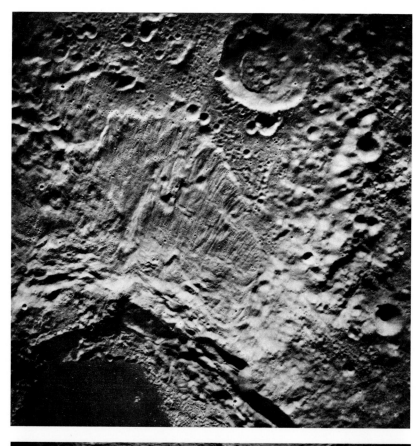

A hand-held colour photograph of the region near the crater Aristarchus (the bright fresh crater at top left), acquired by the Apollo 15 astronauts, shows another part of Oceanus Procellarum. North is to the right in this view which covers an area whose scale can be judged by noting that Aristarchus is about 45 km in diameter. The uplifted Montes Harbinger at the bottom right are thought to be part of the outermost rim structure of the giant Imbrium Basin centred about 900 km to the northeast of Aristarchus. There are numerous sinuous depressions, called *rilles*, in the region: they are probably valleys formed by the flow of molten lava and later modified by the slumping of their walls.

This is an Apollo 15 picture of the same general area as in the previous figure and it includes the craters Aristarchus (left) and Herodotus. An excellent view is provided of the sinuous depression named Schroter's Valley. The picture was taken looking southward across the Aristarchus Plateau onto the level *mare* region Oceanus Procellarum. Schroter's Valley is believed, on the basis of terrestrial analogues, to be a channel, or *rille,* created by the flow downhill of lava that originated in the circular depression between the two impact craters. The distance from this source region to the end of the rille is about 175 km.

Earth and Moon

Early in its history the entire lunar surface is thought to have looked like the highland (or *terra)* region shown in this oblique far side view of an area to the northeast of the Tsiolkovsky Basin. The lunar far side has changed little since the tapering off of the early, massive meteoritic bombardment and the large-scale volcanic flooding of the near side basins did not occur until many hundreds of millions of years later. In this region the meaning of *saturation cratering* is made graphically clear. The craters visible in this picture range in size from a few tens of metres (the limit of resolution) to about 75 km, the size of the large degraded crater in the lower middle of the frame.

As comparison with the preceding picture shows, a typical highland region on the near side of the Moon is much less heavily cratered and generally more subdued than one on the far side. The difference is attributed to the accumulation, on the near side highlands, of blankets of material ejected from the many large impact basins located on the near side. This picture was taken from the Apollo 16 spacecraft looking due south to the horizon and shows the region just south of the Apollo 16 landing site. The caduceus-like object intruding into the picture is the boom of the gamma-ray spectrometer used to map the concentration of chemical elements in the lunar surface (gamma rays are emitted from the surface both as a result of naturally occurring radioactivity and also as a result of radioactivity induced by cosmic ray bombardment). The boom is pointing at the rim of the highly subdued crater Descartes which is roughly in the middle of the frame and is about 50 km in diameter.

The Apollo 17 lunar module landed in the region between Mare Serenitatis and Mare Crisium, which is a mixture of highland and *mare* terrain. The landing site is marked by an arrow in this picture taken from the Apollo 17 service module. The low solar illumination of this view enhances the relief in the region which is relatively much more rugged than that of the earlier landing sites. The region here, where partial volcanic flooding of a highland region has taken place, resembles in some respects the region northeast of the Hellas Basin on Mars. The lunar scene here is about 150 km across.

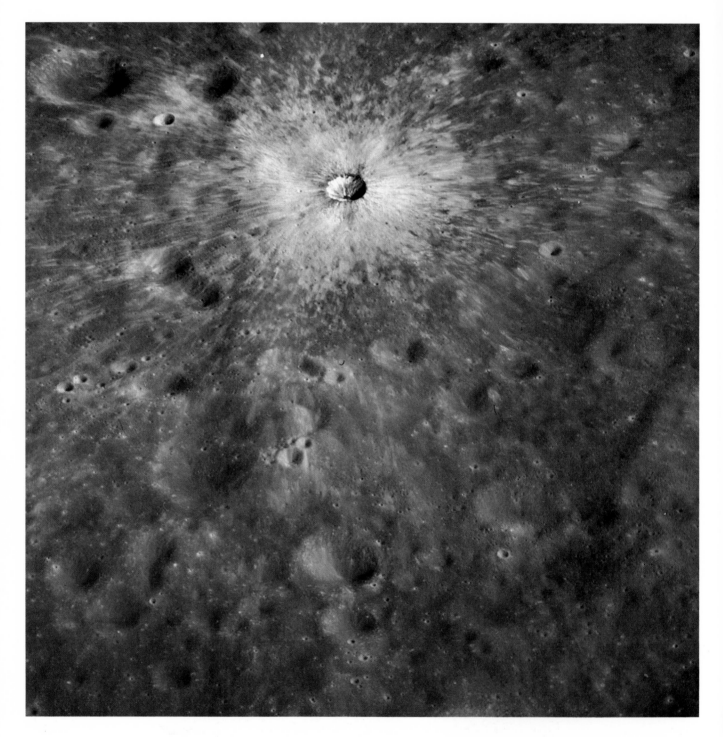

This photograph of a small crater (about 3 km in diameter) on the far side of the Moon provides an excellent example of the bright rays that surround many young lunar craters. The bright material forms an almost continuous blanket around the crater for about two crater diameters and was evidently deposited by material ejected from the crater at the time of the original impact event. Beyond the continuous blanket there is a well-defined radial pattern of rays – streaks of bright material on the surface. With time and the continuing impact of micrometeorites, the bright ray material is 'gardened' into the underlying soil (or *regolith*) so that the rays gradually disappear. The freshness of the rays surrounding this unnamed crater is evidence of its youthfulness.

These Apollo 17 photographs were taken in the southern part of Mare Imbrium on the lunar near side. North is to the top and the prominent crater is Euler, about 27 km in diameter, seen in more detail in the vertical view (bottom). Euler is an excellent example of a young medium-sized lunar crater: it has a sharply defined rim and a well-preserved ejecta blanket extending out a half crater diameter beyond the rim. Concentric terraces formed by the slumping of the walls are visible within the crater as is a central peak within a generally flat floor. In the region surrounding Euler can be seen chains of irregular secondary craters produced by the impact of debris thrown out of a large primary crater. In this case the primary crater is believed to be Copernicus, about 400 km to the southeast – the direction in which the crater chains converge. Copernicus is the only large crater in that direction that is young enough to have created the secondary craters in the *mare* terrain of Imbrium.

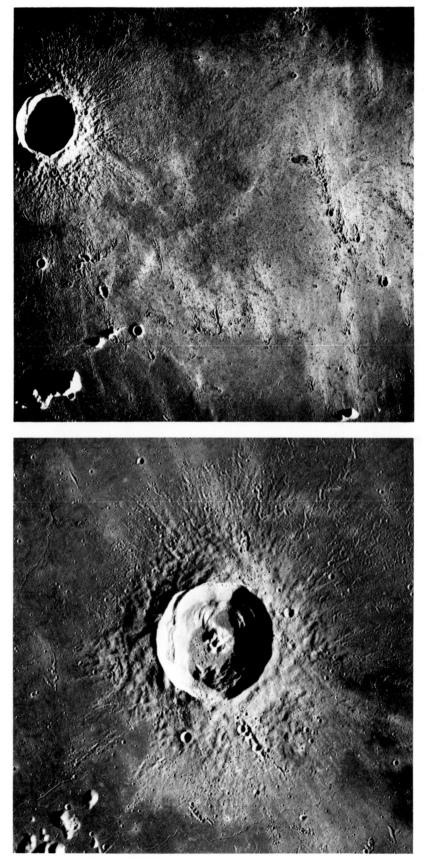

Copernicus is the type example of a large (96 km diameter), young lunar crater. This oblique view was taken from the Apollo 17 spacecraft looking southward over the Carpathian Mountains towards Copernicus. Another, earlier, highly oblique view of the mountainous central peaks of Copernicus, taken by Lunar Orbiter 2 had become internationally well known for the dramatic view that it provided of the lunar landscape, a landscape revealed fully for the first time by the Lunar Orbiters in 1966 and 1967. Copernicus is located on the near side of the Moon less than 200 km from the southern rim of the great Imbrium Basin. It is relatively youthful and is characterized by terraced walls (3–4 km above the crater floor), a rugged but relatively flat floor, and kilometre high central peaks. Rays from Copernicus extend over a large area.

Hadley Rille, located in the area between Mare Imbrium and Mare Serenitatis, was the landing site for the Apollo 15 mission and this picture was taken using a hand-held camera by one of the astronauts on that mission. The frame covers an area of about 17 km by 17 km: the landing site is located at the bottom right-hand side. Data and samples collected by the astronauts indicate that the hypothesized origin of the rille as a lava channel is probably correct. The rille is about 1.5 km wide and 300 m deep.

The landing site of the last Apollo mission was in a valley among the Taurus–Littrow hills on the southeastern rim of Mare Serenitatis. The site was chosen for its potential – judged on the basis of geological mapping using orbital photography – to provide both youthful and ancient rocks. Two of the Apollo 17 astronauts explored the valley with the aid of an electrically powered car. This picture shows one of the astronauts inspecting a huge boulder that has rolled down the side of an adjacent hill, a distance of over a kilometre. Although analysis of the returned samples revealed ages for the basaltic lava that makes up the floor of the Taurus–Littrow valley little different from those of the Apollo 11 samples collected in Mare Tranquillitatis, Apollo 17 was notable in many other scientific respects.

This is a geological map of the Moon, based to a large extent on Lunar Orbiter IV photographs showing topographic forms and textures. The key is not given, since this would require extensive explanation, but the map does show the degree to which the evolution of the Moon is understood. The colour code divides the surface of the Moon into mare, circumbasin, terra plain and crater materials of different ages. For example, the extensive areas of green represent the mare regions, and the yellow shows relatively new craters. A detail from the lower right Mare Nectaris area is shown on the facing page.

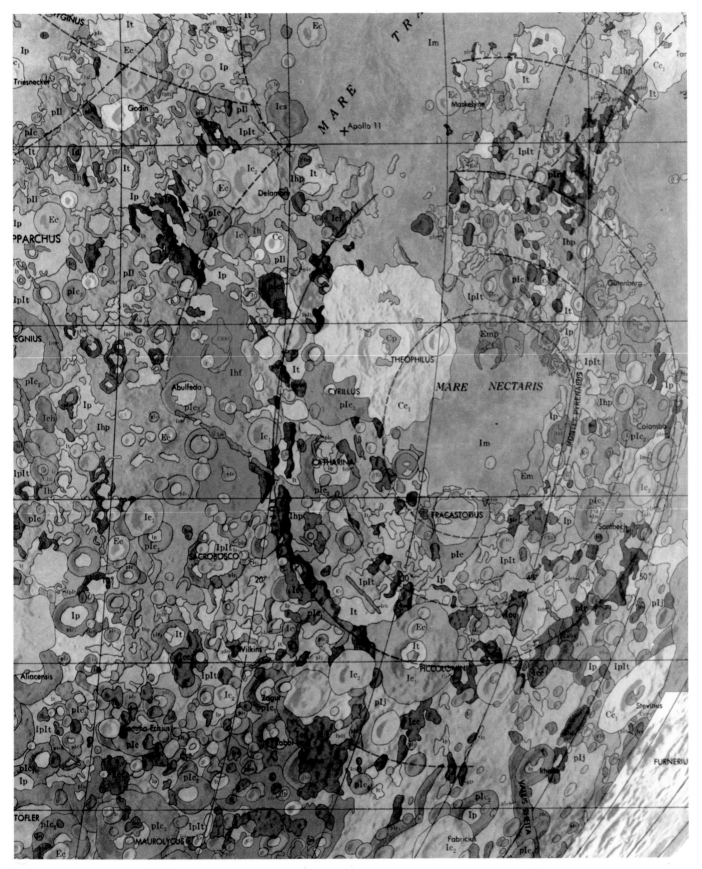

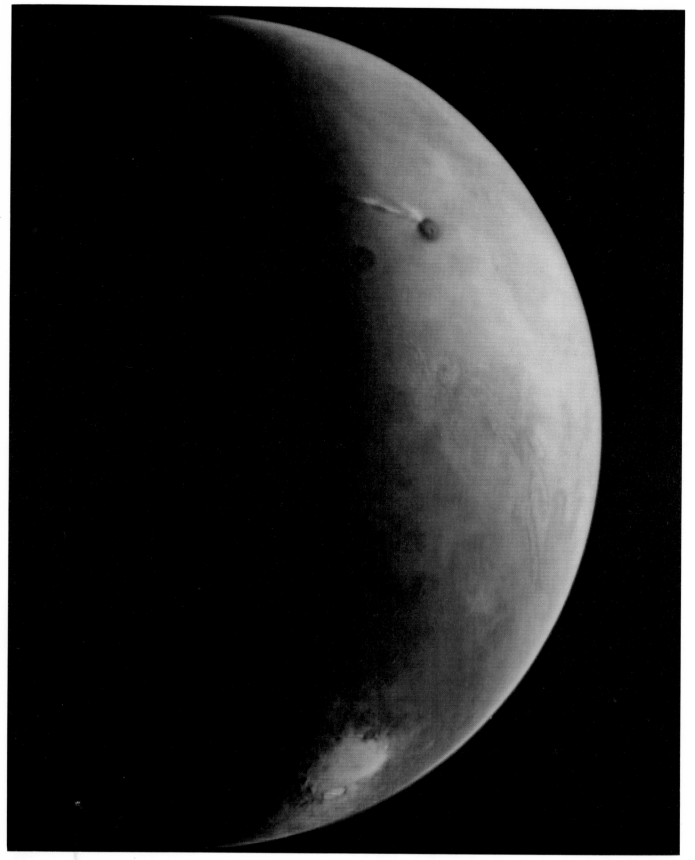

Mars

For centuries Mars has been favoured above the other planets as an object for both serious study and fanciful speculation. Now, nearly two decades since the first spacecraft flew by the planet, there is a substantial body of data concerning Mars that serves to document in some detail the planet's characteristics but which has failed to dispel its mystery or to check speculation. Our fascination with Mars is doubtless due to the relative ease with which the changing face of the planet, the closest body beyond the Earth–Moon system, can be observed and also to the points of similarity with our own world. Mars is the only other planet with both an atmosphere and with surface temperatures that are comparable with (though distinctly lower than) the Earth's. As such, Mars appears uniquely promising as a planet on which to search for life, past or present.

Even before astronomers were able to make definite spectroscopic measurements, the presence of a significant Martian atmosphere was clearly established by visual observations through telescopes. The atmosphere is generally transparent and the surface exhibits continental-scale variations in reflectivity, or *albedo*, that can be seen readily using a moderately sized telescope. These bright and dark markings show a generally fixed pattern. They were mapped, and named, in the greatest detail (a practice which was extended, by some, to imaginary canals). The markings were found to show both seasonal and long-term changes for which numerous explanations were devised, all of which depended to some extent on the presence of an atmosphere. In addition, at times the surface features are temporarily obscured by diffuse white and yellow brightenings that were correctly interpreted to be condensate and dust clouds respectively. Most striking, however, are the white polar caps which grow and retreat with the seasons and, incidentally, greatly add to the beauty of telescopic images of the Red Planet.

Successive spacecraft missions to Mars have revealed much about the diverse and intriguing processes that have been at work on the planet and, in many instances, have confirmed the acuity of the patient telescopic observers of this body. However, partly because the U.S. missions have had mixed objectives, both planetological and biological, there are a number of remote and *in situ* measurements of geophysical and geochemical importance that remain to be made. Significant gaps, therefore, are found in our knowledge of Mars and it is necessary to turn to rather insecurely based models to understand how the planet may have evolved.

Selected geophysical characteristics of the planet, as determined from both Earth-based and spacecraft observations, are summarized in the following table:

Mass 6.42×10^{23} kg	Moment of inertia factor $C/MR^2 \sim 0.365$
Radii† $a = 3399.2$ km ($105°$W)	Day (sidereal) 24 h 37 min 22 s
$\qquad b = 3394.1$ km	Year 687 days
$\qquad c = 3376.7$ km	Obliquity $23° 59'$
Mean density 3.93 g/cm³	Orbital eccentricity 0.093
Equatorial gravity 370.6 cm/s²	

† For a tri-axial ellipsoid of polar radius, c.

Mars is seen to be a relatively small body. However, lacking oceans like those of the Earth, its surface area is comparable to the terrestrial land mass. Mars orbits around the Sun about half as often as the Earth but it has a day and an obliquity that are very close to the terrestrial values. The Martian seasons are, therefore, similar to our own, but almost twice as long. In reference to the seasons it is worth noting that, because of the significant eccentricity of the Martian orbit, the solar heating of Mars is about 40% greater when the planet is closest to the Sun (perihelion) than when it is furthest (aphelion). As will be noted later this variation of the insolation has important consequences for Martian weather.

Our present understanding of the internal structure of Mars is based on theoretical models that have only a few firm constraints. We do not know from direct observation whether or not Mars has a core. The first spacecraft to fly by Mars showed that the planet has, at best, a very weak magnetic field. The spacecraft, Mariner 4, encountered a magnetospheric 'bow shock' and analysis of the data placed an upper limit for the field at about 3×10^{-4} times that of the Earth's. It is unclear whether the observed field is intrinsic to Mars, and an indicator of a core, or whether the field is induced by the solar wind. The results of Russian orbital magnetometer experiments confirm the low level of the field but fail to resolve this question.

By measuring the gravity field of a planet – this is done by observing how the trajectory of an orbiting spacecraft is deflected – it is possible to make inferences about the internal distribution of mass. Certain assumptions must also be made; namely, that the planet adjusts to loads in the manner of a fluid rather than as a rigid body. This is not entirely realistic for Mars and, therefore, one must be cautious about the conclusion that has been reached on the basis of Mariner 9 data – namely, that Mars has a dense core. Presumably such a core would be composed of iron or iron sulphide.

Mariner 9 measurements of the gravity field and figure of Mars revealed an unexpected difference between the northern and southern hemispheres. Mars, like the Earth, is pear-shaped; but, whereas on

Earth the hemispheric difference is measured in tens of metres, for Mars the difference amounts to one or two kilometres (with the southern hemisphere being at higher elevations). Using available gravity and figure data it has been concluded that the lower northern hemisphere has a significantly thinner crust than that in the south. The origin of this asymmetry is unknown; it is assumed to date back to the earliest stage of planet formation. An equally striking departure from sphericity is the presence of a ten-kilometre-high dome, several thousand kilometres in radius, centred equatorially near the 100° W meridian. This region is called *Tharsis* on both old and new maps of Mars, and is the principal centre of volcanism on that planet. Three giant shield volcanoes, the *Tharsis Montes,* and, at the northwestern edge of the dome, an even larger shield, *Olympus Mons,* dominate the area. The Tharsis Montes are aligned in a northeast–southwest direction across the equator on the highest part of the dome, from where they stretch up a further 19 kilometres to their summits, each of which bears a caldera some tens of kilometres across. The precise diameter of each of the volcanoes, several hundreds of kilometres, is hard to define. Individual lava flows on the flanks of the volcanoes can be traced, in certain cases, for hundreds of kilometres. Lavas able to flow for such distances must have been highly fluid; on this basis it is assumed that the Martian shields, like those on the Earth, and like the lunar *maria*, are composed of basalts.

These large volcanic piles might have caused the underlying crust to have been considerably distorted by their weight. However, gravity data derived from tracking orbiting spacecraft suggest that this is not the case – the volcanoes appear to be resting on a largely uncompensated crust. This crust must be very much thicker than the Earth's in order to have the necessary strength. Estimates of thickness vary; the required thickness is probably about 200 kilometres (compared to about 30 kilometres for the terrestrial continental crust).

The considerable loading of the Tharsis dome is expected to lead to a certain amount of seismic activity. Of the two Viking seismometers only one has provided data because the other failed to release from its locked position (the position in which it was set to withstand the shocks of landing). The one working seismometer was limited in its ability to determine the seismicity of Mars (the landing site is on the opposite side of Mars from Tharsis) or the interior structure. A global network of more sensitive instruments will be required before definitive data of this nature are acquired. Nevertheless, the limited data in hand suggest that Mars is much less seismic than the Earth. Seismic data also indicate that the planet is less resonant than the Moon to seismic activity, being more like the Earth in this regard. Present interpretations ascribe this characteristic to the presence of water in the Martian crust, something not found on the Moon.

The asymmetry in the Martian figure accompanies a striking difference in the terrain types that typify the northern and southern hemispheres, as can be seen even from a casual look at the map of Mars. Closer study shows that the actual boundary is not equatorial but, rather, is roughly a great circle inclined at about 35° to the equator and having a maximum northern excursion at about 330°W. South of the division the Martian surface is heavily cratered and contains several large multi-ringed basins, most notably *Hellas* and *Argyre*. This terrain resembles the lunar uplands, but there has been some volcanic flooding and substantial gradation so that the intercrater plains are relatively smooth and the craters themselves are shallow. The first three spacecraft to fly by Mars returned pictures which sampled this type of terrain almost exclusively and gave investigators the impression that the planet had experienced only a limited, lunar-like evolution early in its history.

The mistake was rectified when high-resolution images of the whole planet were returned by Mariner 9 from orbit about Mars in 1971–2. This survey mission discovered a considerable variety of unsuspected surface features including immense volcanoes, canyons, channels and sedimentary terrains. The global Mariner 9 imaging provided the data from which the first high-quality Martian topographic maps were made and showed that in the north the predominant unit consists of sparsely cratered terrain that evidently post-dates the heavily cratered terrain to the south. These plains are heterogeneous, as the Viking orbiter imaging data most clearly demonstrate, and the land forms are among the least well understood on Mars. The division between the major terrain units is a sloping margin where the ancient cratered terrain appears to be in the process of breakup and decay. The remains of the older unit in this margin are often left as *massifs*. These appear to be suffering erosion by mass wasting, with the debris being distributed onto the topographically lower plains region into which this blocky terrain merges. Elsewhere along the margin, and within the southern hemisphere, the ancient cratered terrain has been destroyed by large-scale slumping and breakup of areas to produce disorganized, or 'chaotic', blocky units. Often the areas of breakup form closed depressions and some are obviously associated with broad, sinuous channel-like features which resemble dried-up stream beds. In some cases these *arroyo*-like features emerge 'full bore' from the chaotic terrain and cut a swath for hundreds of kilometres before petering out.

Volcanic activity on Mars is widely distributed, with volcanoes and lava plains found at all latitudes. There are, however, three principal centres of such activity. The centre in Tharsis has already been mentioned. The great height of the volcanoes in this region suggests activity over long periods of time. (In principle, measurements of crater densities could help in pinning down the age of these volcanoes but, unfortunately, ages derived by different investigators, using different models, range from around a hundred million years to two billion.) Also contributing to the great heights of the volcanoes is an apparent lack of crustal motions on Mars, so that a mantle plume of molten rock anchored under a given region could continue to build a mountain for as long as the mantle convection persists and for as long as sufficient hydrostatic head exists to pump the molten rock to the top of the volcanoes. By taking into account the low Martian gravity and making some assumptions about the density difference between magma and crustal rocks, it has been estimated that a crustal thickness of about 200 *Crust* kilometres is needed to provide the hydrostatic pressure for the Tharsis Montes and Olympus Mons (this is consistent with estimates of crustal thickness described earlier).

A second major region of volcanism is in the region known classically as *Elysium*. The two large shields here are located within the northerly plains hemisphere. These are significantly smaller than those in Tharsis and the surrounding lava plains cover an area about one-fourth as great. Although Elysium is generally elevated by several kilometres relative to the Mars mean, the region shows no anomalous gravity signatures at the scales that can be studied using Mariner 9 data.

The third principal region of volcanism is less well defined, lying within the ancient cratered terrain in the general vicinity of the Hellas Basin. Here several volcanic structures are found that have central depressions and a radial pattern of grooves, each tens to hundreds of kilometres long, that clearly distinguishes them from impact craters and, at the same time, from the northern shields. From studies of the areal density of impact craters and from their greater degree of erosion these volcanoes are inferred to be much older than the Tharsis and Elysium shields. They have much shallower profiles and have morphologies that are unique to these ancient Martian constructs. It has been proposed that one reason for the low elevations of their summits is related to their age and that the crust, at the time of this episode of volcanism, was relatively thin so that insufficient hydrostatic pressure was available to build large mountains.

Volcanic activity on Mars has probably been in progress since shortly after the formation of the planet and there is no good reason to suppose that such activity has entirely ceased in the present epoch.

Besides the impact cratering and the volcanoes the most striking surface feature is the immense complex of canyons which dominates the equatorial region to the east of the Tharsis Montes. These canyons, named the *Valles Marineris*, have cut into the Martian crust for several kilometres and, oriented east–west, cross several thousand kilometres of the surface at about 10–15° south of the equator. At their western end the Valles Marineris join a maze of intersecting canyons known as *Labyrinthus Noctis* while, to the east, they merge into the so-called chaotic collapse terrain from which a number of the larger Martian channels debouch. It is presumed that this remarkable canyon, which in places reaches a width of well over 100 kilometres, represents an expression of large-scale tectonism. Plausibly the canyons were formed initially by the pulling apart of the crust to create parallel faults, between which the crustal material began to subside. Continued sinking of this segment of the crust, coupled with erosion of the walls by land slides, has apparently allowed the canyons to develop to the advanced stage we now see. The western half of the canyon complex cuts through moderately cratered plains units that, in some regions, are evidently of volcanic origin and pre-date the Tharsis volcanics. To the east the canyon crosses ancient cratered terrain. Dating the time at which the initial faulting commenced and understanding the nature of the internal evolution that led to this tectonic activity remain important areas of investigation.

Of similar importance is the elucidation of the processes that led to the formation of the chaotic terrain at the eastern end of the Valles Marineris and the accompanying creation of the broad channels that 'flow' out of this area. Generally it is assumed that the channels were cut by water which was mobilized from permafrost and/or liquid subsurface water. Most theories of planetary formation lead to the expectation that Mars would have accreted more water (per unit mass) than the Earth and, although there is reason to believe Mars has outgassed less than our planet, the possibility of there being large quantities of frozen and liquid water in the crust must be seriously entertained.

The broad, flood-like channels on Mars show evidence of liquid flow in their sinuous braiding, but they appear to have been formed by catastrophic events and do not necessarily imply that Mars ever had a climate much different than at present. There are, however, other features that do suggest that at one time the Martian climate may have been warmer and able to maintain liquid water at the surface. These features are dendritic networks of sinuous 'furrows' (tens and hundreds of kilometres long) found throughout the ancient cratered terrain. They resemble patterns of streams that, on Earth, drain regions of elevated topography. If such drainage has occurred then the water may have been distributed in the atmosphere, falling as rain or snow to the surface. Presumably, since the furrows appear to be found almost exclusively in the oldest unit, the postulated era of moderate climate must have occurred early in Martian evolution. Perhaps a different atmospheric composition, with increased amounts of methane and water vapour, could have provided the necessary 'greenhouse' effect to achieve the additional surface heating.

The Martian volcanoes, canyons and channels were all discoveries of the Mariner 9 mission. An additional unanticipated geological unit discovered on that mission was the layered sedimentary terrain of the south polar region. The Viking orbiters have shown that the same kind of surface is found at the north pole too. These polar units are the youngest on Mars, being virtually unmarked by craters, and are superimposed on the cratered terrain in the south and on the plains units in the north. It appears that material eroded from the surface at low and middle latitudes is transported towards the poles where it is redeposited. As a result there is evidence of mantling of the underlying terrains at the higher latitudes that becomes fairly complete poleward of about 75°. It has been estimated that this blanket may be some hundreds of metres thick and in places it has been eroded to form irregular depressions informally known as 'etch pits'. Closer to the poles, and overlying this blanket, are further deposits that have evidently been emplaced in layers. In places the terrain has been eroded to form scarps and hollows where the layers outcrop. Each layer is some tens of metres thick. They form, in totality, a unit that may be some kilometres in thickness. In the north, but apparently not the south, the layered terrain is almost completely encircled by a sea of dunes that may, perhaps, be formed from sand-sized material eroded from that terrain. Alternatively, the finer-sized material transported to the pole in the atmosphere may have been deposited in the layers while larger-sized particles, moving poleward in a different mode, may have been incorporated in the dunes.

It is the presence of the layers, apparently quite regular, that is so provocative. It is difficult to think of any process other than periodic climate change that could have produced layering of the sedimentary polar deposits. The channel features also suggest a once-different climate, and there have been efforts to connect the two observations, with the consequent speculation that Mars alternates between having a cold thin atmosphere (as at present) and a warmer thicker atmosphere able to support running water. It was supposed that the additional atmosphere might be released from the 'permanent' polar caps at times when the orbital characteristics of the planet were significantly different. Such periodic orbital changes, analogous to similar terrestrial changes, are indeed believed to occur and to be of considerable magnitude. However, there now seems to be little chance that the polar caps are thick enough to supply the necessary material.

Mars has two kinds of polar caps, seasonal and perennial. The seasonal caps are known to be composed of carbon dioxide ice, the same compound as the principal (95%) constituent of the atmosphere. Mars is not like the Earth which has oceans and a relatively thick atmosphere to redistribute heat in large quantities about the surface of the planet. On Mars, therefore, the seasonal temperature excursions of the surface follow the Sun's heating rather closely. Near the poles in winter, when the Sun is below the horizon for months on end, the surface temperature plunges and would reach extremely low values indeed if the atmosphere were composed of a less readily condensible gas than CO_2. In fact the surface temperature drop is halted when it reaches about $-125°C$, the condensation temperature of the CO_2 atmosphere at the mean surface pressure of 6 mb. Latent heat release then maintains the surface temperature as the atmosphere condenses on the surface. As a result of this seasonal phenomenon, it was expected that the atmospheric pressure of Mars would vary throughout the year and such a periodic change, which amounts to about 30%, has been observed by the Viking landers.

The seasonal caps grow in autumn and in the first half of winter, reaching latitudes of 40–50° in the south and 50–60° in the north (where autumn and winter occur when Mars is relatively close to the Sun in its elliptical orbit). The caps grow under a cloud 'hood' of water ice and CO_2. This hood largely clears in early winter when all condensed water has been precipitated from the atmosphere leaving only tiny quantities of water vapour. In the middle of winter (when the Sun is moving higher in the sky) the outer region of the cap begins to sublime and a renewal of cloudiness occurs. Winds at the edge of the cap are severe at this time and local dust storms are common. The retreat of the CO_2 caps in late winter and spring takes place gradually and is complete shortly after the summer solstice (the beginning of summer). The polar caps do not entirely disappear, however; relatively small perennial caps remain. The northern cap occupies most of the area north of 80°N while the smaller southern cap is offset from the geometric pole, being centred at about 87° S. Measurements of a $-70°C$ temperature for the condensate material left at the north pole in summer clearly indicate that the ice is water and not CO_2. Thermal balance calculations indicate that the frost has a relatively low reflectivity suggesting that there must be some dirt mixed in the ice. Curiously, the Viking orbiter observations of the southern perennial cap show that it does not experience the same rise in temperature during the summer; it seems likely that the southern perennial cap is composed of CO_2. It has been suggested that the dust-filled atmosphere, typical of the southern summer, may reduce the solar irradiation of the southern cap, leading to the observed asymmetry.

The question of how much water and other volatile substances Mars has outgassed is central to the problem of the planet's evolution. Much speculation has taken place on this subject and the Viking atmospheric analyses have finally provided some data that set major constraints on the problem. Prior to the Viking mission it was known, from telescopic and from orbital spectroscopic and radio-occultation data, that the Martian atmospheric pressure was about 6 mb and that the major constituent was CO_2. Traces of water vapour, oxygen, ozone and carbon monoxide had been measured, but no nitrogen. Measurements made using sensitive mass spectrometers at high altitude during atmospheric entry, and on the surface after landing, show that 95% of the atmosphere is CO_2, ∼2% nitrogen and 1–2% argon. Attempts have been made to estimate the total outgassing experienced by Mars based on the Viking measurement of the abundance of the two nitrogen isotopes, ^{14}N and ^{15}N, in the Martian atmosphere: the measurement shows that the heavier isotope ^{15}N is enriched by 75% relative to ^{14}N when compared with the terrestrial case.

It is highly unlikely that the $^{15}N/^{14}N$ ratio differed from that of the Earth at the time of accretion and the enrichment is therefore viewed as the result of the selective loss of the lighter isotope. In the lower atmosphere of Mars, which like the Earth's is well mixed, there is no obvious mechanism to provide fractionation. In the upper atmosphere, however, the gases are not mixed above a level known as the turbopause and here the concentration of a gas is determined by diffusion. Thus there is an increasing proportion of the lighter gases with height and a means whereby, if gases are lost from the atmosphere in significant amounts, heavier isotopes can become enriched. For Mars, exospheric loss is principally caused by the photoionization of atoms and molecules. The kinetic energy transferred to the atoms is sufficient to let them reach escape velocity and, if the direction of motion is right, loss can occur. The mechanism is complex, but can be modelled, so that the observations can be used to determine the total amount of nitrogen that would be required in the atmosphere to produce the measured enrichment. The results depend on the assumed time of outgassing: the longer the nitrogen is in the atmosphere the less is needed to provide the observed enrichment. This type of analysis indicates that about at least twenty times the present amount of nitrogen would be required, about 2 mb, if the outgassing had occurred early and as much as 30 mb if some of the nitrogen were to interact with the surface to form minerals. The amount of water outgassed is established to be equivalent to 120 m averaged over the planet.

What do the results of the Viking surface elemental analysis tell us about the evolution of Mars? At two sites, both within the northern hemisphere and separated by many thousands of kilometres, the Viking landers subjected samples of Martian soil to analysis by an X-ray fluorescence spectrometer sensitive to elements whose nuclei contain more than 12 protons (many important, light elements were not detectable by the instrument, including hydrogen, oxygen, carbon and sodium). Analyses of numerous samples at both sites showed considerable similarity and suggest that the fine materials on Mars are rather homogeneous, presumably as a result of planet-wide redistribution at times of major dust storms.

The results also show that the abundances of iron, magnesium and calcium are high while those of aluminium, silicon and potassium are low. The material is closer in composition to the magnesium- and iron-rich (*mafic*) silicates that make up the terrestrial mantle than to the lighter aluminium ('sialic') silicates making up the Earth's crust.

The high content of sulphur in the Martian fines is notable and, as yet, unexplained. The surfaces sampled by the Viking landers show a surprising cohesion as if the fine material were cemented together; magnesium sulphate has been suggested as a possible bonding agent.

Although the Viking spacecraft did not carry any instruments able to identify minerals, as such, attempts have been made to understand the elemental abundances in terms of possible mineral assemblages. No plausible mixture of primary igneous rocks appears to satisfy the results, suggesting that the fines are weathering products. One good fit is provided by a mixture of clays together with smaller amounts of other minerals.

Although there are considerable uncertainties about the mineralogy of the Martian fines, and there are no elemental data for any rocks, it is clear that the analysed material is derived from magnesium-rich and iron-rich silicate rocks. It is unlikely that there are any alkali-rich granitic minerals exposed over significant areas; this argues that the Martian crust is substantially less differentiated than the Earth's. It is generally believed that Mars has a core, but the secondary differentiation of the lighter silicates from the ferromagnesium silicates of the mantle appears to have been limited. The indication that the Martian mantle and crust are relatively undifferentiated is in line with the analyses of atmospheric composition that imply limited degassing. Orbital imaging indicates that Mars has no obvious signs of plate tectonism and the degree of volcanism on Mars, though spectacular, is about two orders of magnitude less in scale than on the Earth. On the

Earth, plate tectonics, volcanism and differentiation are related manifestations of internal convective activity and it seems that Mars has never achieved the necessary scale of internal activity to bring about this degree of evolution. Probably the early formation of a thick, strong crust and the relatively rapid dissipation of heat occasioned by the greater surface-area/volume ratio of Mars are sufficient factors to account for the difference. It is of interest to look at theoretical models of the thermal history of Mars with 'reasonable' assumptions for the early heating due to accretion and for the radioactive elements incorporated into the planet. Without going into the details of how such models are constructed it is instructive to examine the results of one of these. Basic assumptions for this model include the content of radioactive elements. In this model differentiation of a crust occurs early owing to partial melting near the surface as a result of the accretionary heat (due to the infall of material to the surface). This heat is also sufficient for an Fe–FeS core to form early, sustained by the extra gravitational heat resulting from the migration of the heavy liquid to the planet's centre and, less importantly, by the increasing heating of radioactive decay. Core formation is complete by about 1 billion years. The silicate mantle begins to melt in its upper layers after about 2 billion years and this melting works inward. The model predicts that, at present, the planet has a lithosphere about 200 kilometres deep, a mantle still capable of solid-state convection, and a molten core. Such a thermal history would clearly involve extensive internal activity. The discovery that the Martian surface materials are highly mafic is not necessarily in contradiction with such a model. The Martian interior may have undergone considerable melting and consequent differentiation but this may not have led to the same high degree of differentiation at the surface for the reasons previously mentioned. Equally, however, we have only some very general data on which to base thermal models and it is very important that the internal structure of Mars be more directly probed by seismic, gravity and magnetics experiments in order to provide us with a much more complete picture.

What do the other data, principally photogeologic and gravity data, tell us about the evolution of the surface? There are still many pictures of the surface which show processes we understand little about but, in broad terms, the stratigraphic record on Mars suggests that the surface evolution has fallen into five phases. The first begins toward the close of the accretionary episode when the entire surface would have been heavily cratered. The difference between the northern and southern hemispheres is likely to date back to non-uniformities in this early phase when the core formed and crust differentiated from the upper

131

mantle. In the north the crust was evidently thinner and the surface elevations were lower than in the south.

In the second phase the cratered terrain in the north, where the crust was thinner, was broken up under the influence of forces, presumably internal, as yet poorly understood. The radial faults that accompanied the beginnings of the Tharsis dome date back to this second phase and related tectonic activity may have helped to break up the northern crust. The internal activity that produced the major global asymmetry of the Tharsis bulge is less than well understood: substantial internal convection is implied. The bulge may not have been at the equator originally but the rotation axis of Mars may in time have migrated to a direction that achieved this. The erosion of the dendritic 'furrows' in the cratered terrains appears to belong to this early phase and thus implies an early atmosphere thicker and wetter than today. Early degassing associated with accretion and crust formation may have produced a relatively massive atmosphere which, with an increased greenhouse effect due to methane and water vapour, may have been significantly warmer than at present. The erosion due to the 'drainage' of precipitation (if that indeed is what the furrows represent) was evidently limited and the warmer climate may have been of limited duration or the water may have been rather quickly lost to the near-surface interior and to the polar caps. Some of the massive channels found in the equatorial region are also, apparently, of rather early origin and belong to this second phase. Water may have accumulated in low areas with flooding produced when this water was released as a result of meteoritic impacts. Alternatively the floods may have been due to volcanic activity leading to the melting of ice dams that were holding the water in check. A good deal of surface weathering and the production of clay could have occurred in this era.

In the third phase, volcanism, centred in the Tharsis region, led to the inundation of an area many thousands of kilometres across which has since acquired a significant population of craters. This must also have been an important period for planetary outgassing though temperatures probably never allowed liquid water to form.

Phase four is marked by the continued influence of the crustal loading produced by the Tharsis area with accompanying faulting and, in particular, the formation of the rift valley that has since enlarged to form the Valles Marineris.

Since that time, phase five, the principal processes at work on Mars have apparently been continued volcanism and low-latitude erosion due to mass wasting and aeolian (wind) processes. Large quantities of material have been moved poleward throughout the history of Mars and these have accumulated as thick sedimentary deposits near the poles. The layers observed in these deposits may represent major climate changes or, perhaps as likely, relatively minor changes that are sufficient to modify the global winds and the frequency of dust storms.

Violent planet-wide dust storms demonstrate that the thin Martian atmosphere is highly dynamic. Considerable progress has been made in understanding the general circulation and dynamics of the atmosphere in the past 15 years as a result of theoretical analyses and studies based on computer models of planetary circulation. Measurements and pictures acquired by orbiting and landed spacecraft have recently served to confirm much of the previous analytical work and to bring to light interesting new phenomena.

In many respects the dynamics of the Martian atmosphere are simpler than those of the Earth: there are no oceans to complicate the heat transport problem and to provide unlimited amounts of condensible moisture with its important release and absorption of latent heat. On the other hand, the Martian case has unique characteristics for which allowance must be made. For example, the topography on Mars varies by about 30 kilometres and reaches heights where the atmospheric pressure is almost an order of magnitude less than at the surface. Direct heating of the atmosphere by heat reradiated from the surface is very important and, because of the large-scale topography, this has a strong effect on the wind regimes. The atmosphere of Mars is much thinner than the Earth's and has a much smaller heat capacity. It therefore heats up and cools down more rapidly than our own atmosphere, having a 'radiative relaxation time' of about two days, compared to the terrestrial case of about 40 days. This short time-constant leads to a much closer coupling of surface and air temperatures than on Earth. Thus on Mars the atmospheric temperature at the top of one of the towering shield volcanoes will be about the same as the air temperature at the bottom of a deep depression: the atmospheric isotherms accordingly tend to parallel the elevation contours. An elevated region will tend to be associated with a low-pressure region owing to daytime heating, and a topographic depression will tend to produce a high-pressure area. The high-altitude regions of Mars are therefore characterized by daytime upslope winds, and cyclonic motions, and by night-time downslope winds. Basins show an opposite behaviour.

The responsiveness of the Martian atmosphere to heating also shows itself in the significant amplitude of diurnal and semi-diurnal *tides*. Some of the heating of the atmosphere takes place directly as a result of the absorption of solar radiation. This direct heating can be appreciable especially if the atmosphere is dusty, as is often the case (then the sunlight is absorbed directly by the dust particles which can heat the atmosphere efficiently). The daily cycle of heating leads to a marked temperature, and thus pressure, cycle. This planet-wide pressure oscillation is most intense at tropical latitudes, and the associated winds, which show a rotation in direction during the day, increase in strength with height.

The two Viking landers, one of which landed just within the northern tropics (22°N) and the other much further north (48°N), have confirmed that the phenomena discussed above do take place. Viking 1 at a landing site in the Chryse depression returned data showing a daily summertime wind variation consistent with the geostrophic winds expected for a point on the western side of the basin – a light daytime wind from the south associated with anticyclonic air motion around the high and gentle night-time winds from the north. The daily and semi-diurnal tides are clearly seen in the recorded pressure variations and, as predicted, are of lesser amplitude at the more northern site.

The summer hemisphere winds tend to be dominated by topographic effects because there is only a small equator-to-pole temperature variation to drive the general circulation. In fact, at the summer solstice the daily insolation at the illuminated pole is rather greater than at the equator. In the winter hemisphere, however, there is a strong equator-to-pole temperature gradient and, in addition, a major transfer of mass to the pole as the atmosphere condenses in that region. A slow overturning of the atmosphere in a Hadley-cell-like motion is unstable under these conditions and, as on the Earth, a wave regime is established with prevailing westerlies and a mid-latitude jet stream. Heat is transferred poleward by tongues of warm air; cold air moves equatorward in a similar fashion. These travelling disturbances with their associated fronts have been detected by the Viking landers and the extended cloud boundaries that can mark such fronts have been photographed from orbit.

Mid-latitude westerlies are the prevailing winds during autumn and winter when the insolation in that hemisphere is a minimum. Because the seasonal polar cap is present throughout spring the strong equator-to-pole temperature gradient continues through this season also. In fact, because the Sun is now heating the ice-free part of that hemisphere increasingly strongly, there is an extremely large temperature gradient near the margin of the cap. The westerly winds are, therefore, very strong in this region in spring and generation of dust storms is common at that time near the cap.

Because the atmosphere of Mars is quite thin, the wind velocities needed to raise dust are much higher than on Earth. The great planet-wide dust storms occur in the southern spring–summer period which also coincides with the time when Mars is closest to the Sun. The scale of the disturbance suggests that the atmosphere is subject to an instability. An instability can arise because the injection of dust into the atmosphere leads to increased absorption of heat by the atmosphere and a consequent increase in local winds (driven by temperature–pressure differences between dust-laden and clear air) and in planet-wide tidal winds: these increased winds in turn lead to the further injection of dust. Tiny micron-sized particles can be lifted to great heights (Mariner 9 observations suggest 50 or more kilometres) and take a long time to settle out as they are not cleaned out by rainfall as on Earth. Dust raised into the atmosphere by the strong winds near the retreating cap edge can remain in the atmosphere and, as southern summer solstice and perihelion approach together, can prime the atmospheric heat engine which responds with impressive fury.

The Martian orbital elements, in particular the eccentricity and obliquity, are subject to large periodic oscillations. A consequent modulation of the perihelion-solstice forcing of the atmospheric circulation is implied and it seems reasonable to suppose that the intensity and frequency of global dust storms will be correspondingly affected. It may be such a modulation that is mirrored in the sedimentary layers of the polar regions.

The Martian atmosphere contains only a trace of water vapour – rarely more than a few tens of precipitable microns. Nevertheless, at the characteristic temperatures of Mars, the air is never far from saturation. Condensate clouds are consequently a common and attractive feature of the Martian scene. Always present in one hemisphere or the other are clouds associated with the polar regions. The autumn growth of the seasonal CO_2 cap occurs under a general haze of water and, probably, carbon dioxide ices. In early winter most of the water ice appears to fall out and the atmosphere clears greatly. The air near the cap edge has a very stable vertical temperature structure (an inversion) at that time because of the low-level cooling. In these circumstances air deflected upwards by the strong westerlies as they encounter positive topographic features, such as ridges and crater rims, tends to undergo a vertical oscillation in the wake of the obstacle (the rising air cools more rapidly than the surrounding air and therefore sinks, overshooting its original position and becoming warmer by compression so that it rises again). These oscillations resemble the waves that form behind a boulder in a stream. In orbiter images the waves can be seen because water vapour and/or CO_2 condenses at the crest of the wave, as a result of the air cooling as it rises, and regular trains of clouds are formed that may stretch for hundreds of kilometres.

Elsewhere, and especially in elevated regions, the atmospheric temperature profile is generally less stable in the middle of the day and convective clouds can be produced. Such clouds, formed at the top of a discrete convective cell, have been observed in orbital pictures. The cloud heights, about 5–8 km, can be determined by shadow measurements and provide a measure of the height within which the convection is ocurring, a height somewhat greater than on Earth.

Besides the convection, which occurs in a relatively shallow layer, the Martian air is subject to uplift over significantly greater heights as a result of motions induced by the large-scale topography. As might be expected the giant Tharsis shields and Olympus Mons are the sites of condensate cloud formation due to such *orographic* uplift. The clouds associated with the volcanoes can be readily seen from the Earth and each was described as an element of the 'W Cloud', so-called because of the pattern as seen at the telescope where south is at the top of the image (to most viewers it is hard to see the pattern of any letter no matter what orientation is examined). Spring and summer, when the water vapour content of the air is greatest, are the seasons when these orographic clouds form most strongly. They develop slowly in the morning as the upslope winds increase with diurnal heating and during the afternoon hours may cover very large areas, generally to the west of the mountains as a result of the prevailing easterly winds near the equator.

A rather general low-level haziness due to condensed water vapour is typical of the cooler times of day, dawn and dusk, in the summer hemisphere. At night the lower few kilometres of the atmosphere cool greatly when the reradiated heat from the surface falls drastically. Water vapour in this layer condenses to form a fog that can linger for some hours after daybreak. Some of the condensed water vapour may find its way to the surface at night to form a light frost.

Other types of clouds have also been observed from orbit at heights as great as 50 km. Thin haze layers, composed of condensed water vapour in some instances and of CO_2 in the case of the higher layers, are common and also have been observed at heights of many tens of kilometres. A sharp edge to the planet's disc is almost never seen due to the dust and condensates in the atmosphere.

The water vapour in the atmosphere, besides making possible the formation of clouds to enhance the scene, also plays a fundamental role in the photochemistry of the atmosphere. Without the tiny amounts of water vapour it is possible that Mars would not have an atmosphere composed of CO_2. This molecule, unshielded from the solar ultraviolet radiation, is readily broken apart to form CO and O, the recombination of which to maintain a CO_2 atmosphere, is not well understood. Without an efficient recombination mechanism, carbon monoxide and molecular oxygen would become the principal atmospheric constituents. Water vapour is also subject to photodissociation, to form

O and OH, and it is thought that these highly reactive species may serve to catalyse the recombination of CO and O. Another hypothesis is that excess molecular oxygen is broken up and recombined with carbon monoxide, not in the gas phase, but in a solid phase reaction involving the oxidation of ferric iron at the surface.

Other photochemical reactions that occur in the Martian atmosphere produce ozone (O_3) by combining atomic and molecular oxygen, both of which are present in small amounts. Insufficient ozone is created to provide an efficient shield against the solar ultraviolet flux since there is an efficient catalytic reaction, involving the dissociation products of water vapour (O and OH), that destroys the ozone. Since there is very little water vapour in the winter hemisphere of Mars, detectable levels of ozone can, and have been, observed by an orbital ultraviolet spectrometer.

Hydrogen peroxide (H_2O_2) can also be formed in small quantities in the atmosphere and, adsorbed on to the soil particles, may help account for the highly oxidizing nature of the soil discovered by the Viking landers when performing wet chemistry experiments involving samples of Martian soil to look for evidence of life processes. This property of the soil is certainly a major antagonist to any organic molecules that may be formed at the surface. One of the Viking biology experiments demonstrated, somewhat paradoxically, that the soil is capable of reduction also. The results of this experiment, which involved no liquid reagents but simply measured the amount of 'labelled' CO_2 that was incorporated into an acquired soil sample after some days under simulated external conditions, could have been interpreted as evidence of metabolic activity if there had been any indication of the presence of organic molecules in the soil. One of the Viking instruments, the same one that made the analysis of atmospheric composition, had the measurement of organic molecules as its prime objective. A biological synthesis of organic molecules has been demonstrated on Earth for simulated Martian conditions and it was expected, therefore, that some would be present in the soil samples by Viking. Surprisingly there were no organic molecules detected to the parts per billion level. In these circumstances the likelihood of any of the biology experiment results being indicative of biological activity is small. It seems that the Martian environment, with its intense ultraviolet flux and its oxidizing character, is not one favourable to the development of complex organic molecules. The search for life, and in particular for evidence of past life, is by no means ended, but the prospects do not seem bright. Jupiter or Titan, with their reducing atmospheres, appear to be friendly environments for organics but whether they can provide the conditions under which life might develop is highly uncertain.

Olympus Mons, besides being a major geological feature of Mars, is also an important influence on Martian weather. In the summer the volcano is a centre for large-scale cloud development as illustrated here. This Viking Orbiter frame was acquired at a local Mars time of 8 a.m. in early summer and shows how diffuse haze and convective cloud encircles Olympus Mons at that time of day. The volcano is approximately 500 km across and has a complex central caldera which is itself 80 km across. The mountain rises to a height of 25 km above the surrounding plain and the upper levels reach well above the clouds. An escarpment, or cliff, delineates the shield and has a height of up to 6 km.

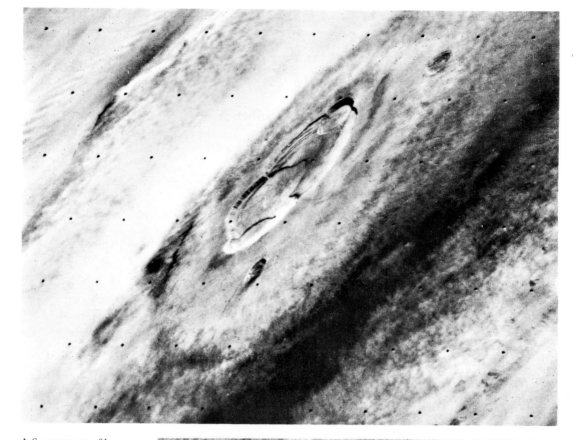

A fine pattern of lava flows is revealed in this close-up view acquired by a Viking Orbiter of the lower southwest flanks of Olympus Mons. Detail is enhanced by the non-uniform distribution of dust that has been carried on to the side of the mountain by strong winds. The picture measures about 100 km on each side.

Ascraeus Mons is the most northerly (12°N, 104°W) of the three volcanoes (Ascraeus, Pavonis and Arsia Montes) aligned along the Tharsis Ridge. All three reach a height of about 25 km above the reference geoid – substantially higher than any mountain on Earth. A conical mountain, Ascraeus Mons differs from the other two in having a complex caldera consisting of several coalescing circular structures, each caused by the collapse, after an eruption, following the withdrawal of the underlying magma. At the edge of the volcano, both to the north and south, are found intricate patterns of irregular lava channels. This oblique picture was taken, looking at the mountain from the south while the Viking Orbiter was at a very high altitude, in early morning on a summer's day: an extensive water ice cloud hides the lower western flanks. Such clouds commonly form at this time of year as a result of air moving up the slopes of the mountain and cooling as it rises.

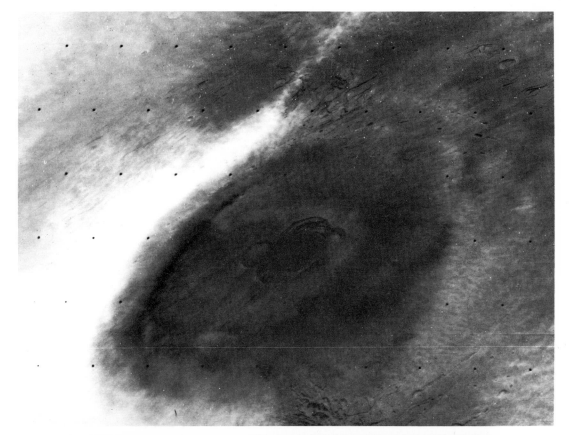

Tharsis is the principal centre of volcanism on Mars. The most southerly of the Tharsis volcanoes is an immense shield named Arsia Mons. This Viking orbiter colour composite of Arsia Mons and the surrounding volcanic plains allows the tracing of individual lava flows for many hundreds of kilometres. Many of the lava flows appear to originate in a *re-entrant* on the southwest flank of the mountain. The appearance of the flows changes with distance from Arsia Mons – broadening and becoming less well-defined with distance: steepness of the slope, changes in lava characteristics, and differences in eruption rate may all be involved.

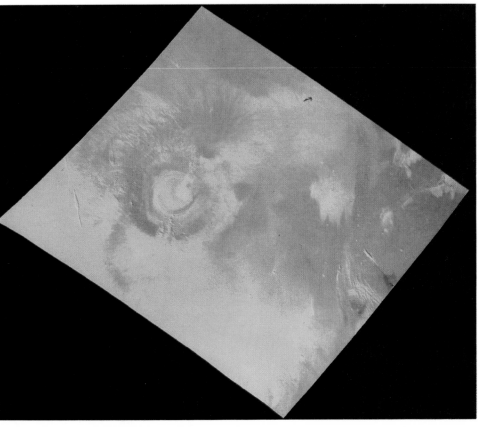

Set in a volcanic plains region within the cratered terrain of the southern hemisphere is another of the class of upturned-saucer volcanoes – Tyrrhenum Patera located at 22°S, 253°W. This feature was formed relatively early in the planet's history judging by the large number of impact craters and by its degraded appearance.

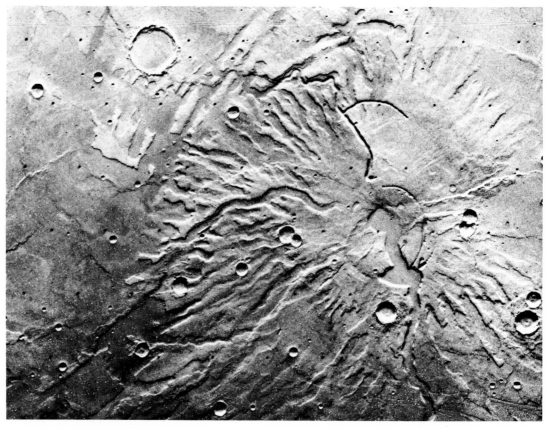

Apollinaris Patera, located at 8°S, 186°W is an excellent example of an ancient central volcano. It is on the margin between old cratered terrain and the plains units of the north. Apollinaris Patera is readily distinguished from the younger Tharsis and Elysium volcanoes both by its relatively shallow profile and by the greater degree of cratering and erosion that it has suffered. The caldera of Apollinaris Patera is about 100 km across and, on the southeast side, lava flows can be traced for a further 200 km.

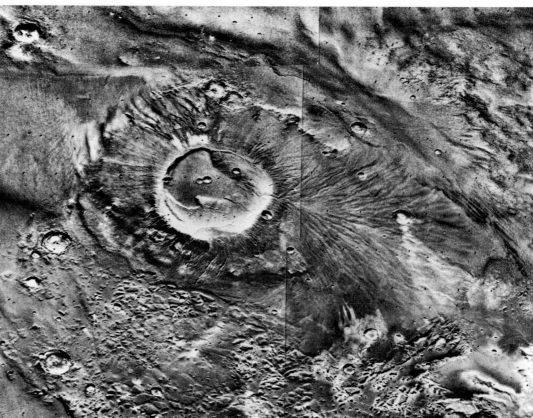

The Nilosyrtis region at 34°N, 290°W lies in the transition between the southernly cratered unit and the northernly plains unit. This Viking Orbiter mosaic covers an area about 150 × 80 kilometres and shows numerous rounded hills between which lie flat-floored valleys. The floors show curvilinear patterns of parallel grooves that run the length of the valleys, fanning out at their ends and suggesting the action of some mechanism of flow. They resemble terrestrial surfaces where the freeze–thaw cycle of water contained in the surface material has led to gradial mass flow downhill. It is not unlikely that there are large amounts of permafrost beneath the Martian surface and some kind of an analogous mechanism may have caused these Martian landforms.

This mosiac of pictures acquired by Viking Orbiter 1 in September 1976 is hardly one of the more beautiful sets returned from Mars. It does, however, provide our best view of the 1500 km diameter Hellas Basin. Generally dust and condensate haze obscure the floor and some are undoubtedly present even on this relatively clear winter day. The contrasting albedo patterns appear to delineate major lava flows that quite probably were precipitated by the cataclysmic impact that caused the 6 km deep basin.

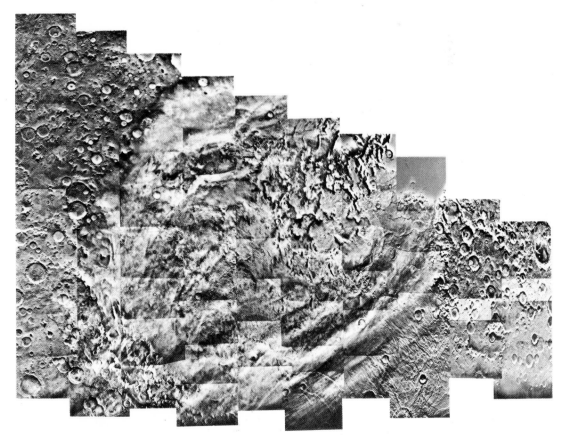

These two views of the Lunae Planum region cover a very large area to the west of the first Viking landing site. The colour composite provides a vertical view of the region in the top left hand corner of the oblique black-and-white picture. The latter frame provides a view from the south and stretches to the limb of the planet, covering about 1500 km in the transverse direction. The irregular escarpment bisecting the picture marks the boundary between the two major terrain classes on Mars – the more elevated, heavily cratered terrain of the southern hemisphere and the lower plains regions of the northern hemisphere. In this region of the planet the boundary runs roughly north–south with the older, higher terrain to the east. The mechanism that is causing the steady disintegration of the more ancient crust is not well understood but is thought to involve sapping processes liberating permafrost.

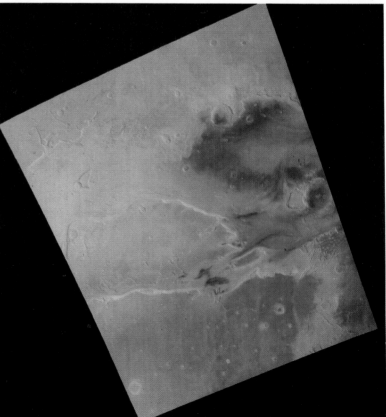

Among the last pictures taken by the Viking 2 spacecraft on approaching Mars was this view of the Hellas Basin. (Special processing has been applied to this picture, made from three black-and-white filtered images, to remove the effects of changing solar illumination and this has led to a bluish terminator and a reddish limb.) This huge basin, 1500 km across, is at least partly covered by the snows of the southern polar cap but there is also evidence of haze, both condensate and dust, which has to date prevented a clear view of the surface there.

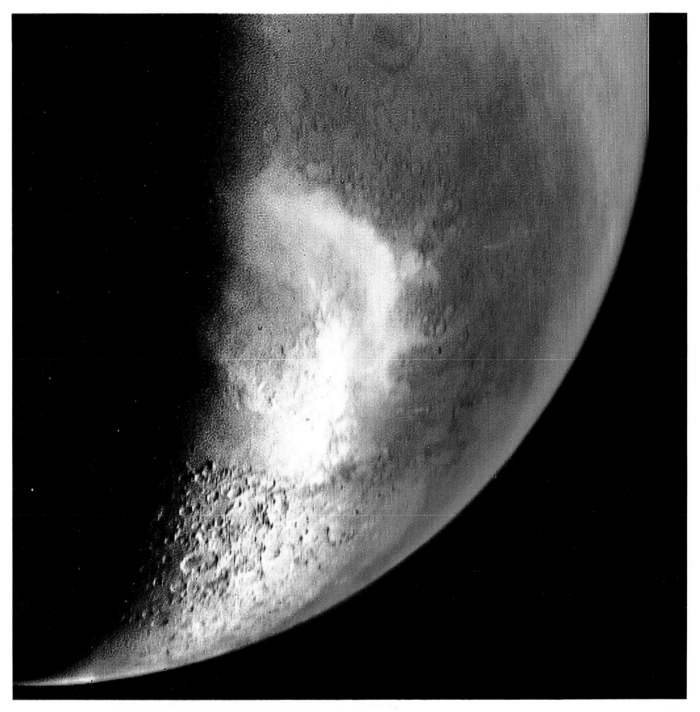

A variety of different crater types are found on
Mars, some of which resemble those on the Moon
and on Mercury. One class of Martian craters
appears to be unique to this planet. Craters of this
class are distinguished by the character of the
material surrounding the crater, material that has
been excavated by the meteoric impact and
redistributed as an 'ejecta blanket' around the
depression. On Mars in many instances the ejecta
seems to be made up of several layers, each one
being demarked by a low ridge at its outer edge. The
example at the top here is a crater named Yuty at
22°N, 34°W and it possess the characteristic layered
ejecta blanket. Each layer has a complex lobed outer
ridge. The depression is about 15 km across and the
region covered in the picture is about 50 km across.
The bottom photograph shows the 30 km diameter
crater Arandas at 43°N, 14°W. The ejecta
configurations suggest that the blankets were
produced by a mechanism involving the flow of
material rather than simple ballistic expulsion from
the evacuation. Mars differs from the Moon and
Mercury in having an atmosphere and, most likely,
large quantities of permafrost. The impact of a large
meteorite into the surface of Mars might readily
lead to the entrainment of atmospheric gases and/or
heated subsurface volatiles into the ejected material.
This might then flow outward *en masse*. The most
extensive ejecta blankets are found at low altitude at
the colder latitudes, a finding consistent with the
idea that subsurface volatiles are involved in the
process by which these 'fluidized' craters were
formed.

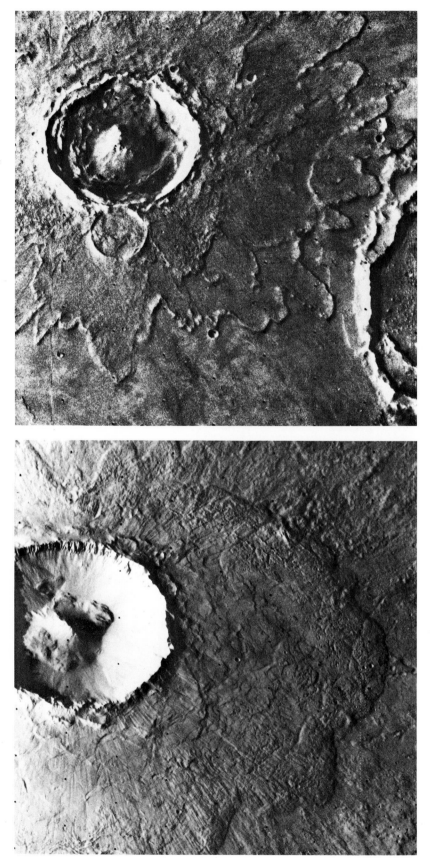

This ghostly scene was recorded in early winter shortly after daybreak and shows a region on the northeastern edge of the great Hellas Basin (the region is adjacent to, and to the west of, the area shown in the following figure). The Hellas Basin, to the left of the frame, is a 1500 km diameter, 6 km deep depression created by an enormous meteoritic impact event early in Martian history. Difficult to observe at all times because of dust and condensate hazes, the floor of the Hellas Basin appears to be largely covered in lava flows. These volcanic floods have not, however, led to a basin interior as uniformly smooth and level as those of the large lunar basins. The larger of the two bright 'ghosts' is a 600 km long channel named Harmakhis Vallis which originates on the outermost southern flanks of an ancient saucer-shaped volcano named Hadriaca Patera. The brightness of the channel itself is thought to be the result of low-level water ice hazes that have formed inside the channel just after dawn.

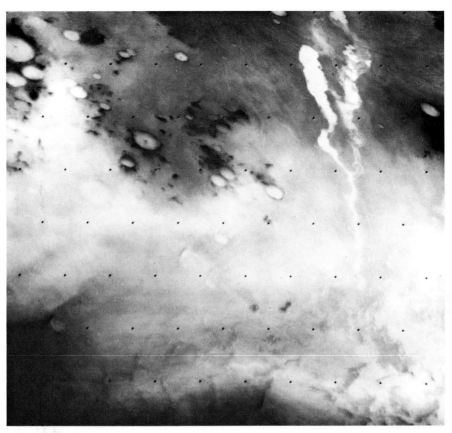

This Viking Orbiter view of a broad area (about 250 km on a side at 35°S, 257°W) to the northeast of the Hellas Basin was acquired early in the mission under conditions of excellent atmospheric clarity. The region resembles certain areas of the Moon where *mare* lavas have flooded the older upland terrain. Here such lava flows have inundated most of the region, covering all but the largest of the ancient craters. Two of the craters are barely visible – marked only by the remains of the craters' rims. In all cases the ejecta material surrounding the original craters has been covered and the interiors have been partially filled.

The immense Martian canyon system, Valles Marineris, was named in honour of the Mariner 9 mission that led to its discovery in 1971. The Mariner Valleys are located just south of the equator to the east of the Tharsis rise and the system of canyons stretches more than a quarter of the way around the planet: in total the canyons cover a distance of about 4500 km and measure from 150 to 700 km in width. The vertical relief varies from about 2 to 7 km. At its western end the Valles Marineris is made up of a complex canyon network named Noctis Labyrinthus. Moving east this network merges into a series of broad, steep-walled canyons that approximately parallel the equator. The relatively uncratered floors of the canyons might speak of a comparatively youthful origin but, on closer examination, it is seen that the floors are covered in landslides which have served to continuously renew the canyon floors. Crater counting is therefore not very helpful in assessing the age of the system. The origin of Valles Marineris is thought to have occurred quite early in the history of Mars and to have been the result of large-scale tectonism. Crustal subsidence was probably important in initiating canyon formation. The subsequent evolution of the canyons probably resulted from further tectonic activity and also from slumping of the walls followed by the transport of the slide material from the floors by wind action.

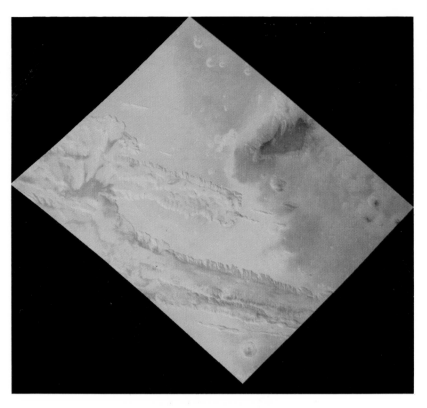

The different sections of Valles Marineris are individually named. Tithonius Chasma, immediately to the east of Noctis Labyrinthus, is where the great Martian canyon system takes on its typical form. Here the canyon is about 50 km across in the narrower part to the right of this Viking Orbiter picture. Data from the earlier Mariner 9 mission indicate that the vertical relief is about 4 km. The steep-sided canyon walls are seen to have a characteristic spur and gully form, apparently under structural control, i.e. the orientation of these features appears to be determined by pre-existing weaknesses in the rock. Considerable detail can be observed in the canyon floor which is covered in debris resulting from landslides that have eroded the walls. Sharply defined lobate flow fronts mark the boundaries of individual slides. Only a few impact craters can be discerned.

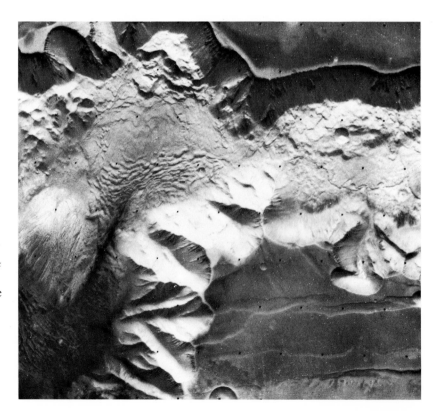

This close-up view is of a 120 km section of the western part of Valles Marineris, named Ius Chasma. The southern wall (bottom) is broken by numerous tributary canyons that typically stretch for about 50 km into the surrounding upland plain. On a smaller scale these tributary canyons resemble the Noctis Labyrinthus complex and, like that part of the great canyon, are thought to have been formed by tectonic processes rather than fluvial action. Running like a spine down the centre of Ius Chasma is a steep, heavily eroded ridge, which appears to reach approximately the level of the surrounding plain. It seems likely that at an earlier time Ius Chasma was made up of two parallel canyons which have subsequently widened until the two canyons merged into one.

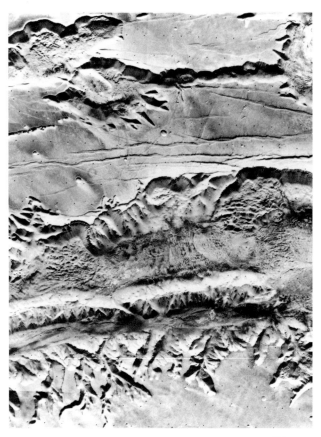

The type of Martian terrain called 'chaotic' was first recognized in pictures taken by the Mariner 6 spacecraft which flew past Mars in mid 1969. Most areas of chaotic terrain are contained within a region of ancient cratered terrain centred near the equator at about 30°W. This surface type is made up of a jumble of blocks that are at a somewhat lower elevation than the surrounding terrain and appears to have been formed by the collapse of that terrain. The mechanism of formation is debated but is generally conceded to be related to the presence of ground ice: numerous channels have their origin in regions of chaotic terrain, suggesting that the channels and the collapsed areas are genetically related. The melting of ground ice, by subsurface magmatic heating or by the impact of a meteoroid, might provide the trigger. This picture, located at 12°N, 56°W, is about 50 km on each side.

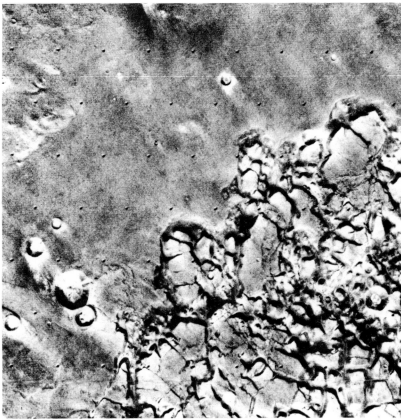

In this Viking Orbiter picture is shown an example of the many networks of dendritic channels that have been cut into the ancient cratered terrain of Mars. In this particular case the channels are made more easily visible by the presence of frost associated with the southern winter cap. This network is located at about 40°S: others have been recognized at comparable latitudes north of the equator. The channel systems resemble stream channels on Earth that result from the runoff of rainwater. It is tempting to ascribe a similar origin to them on Mars but, at present, this is a matter of considerable debate among planetary geologists.

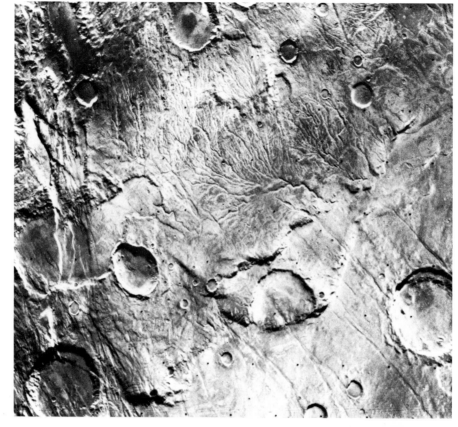

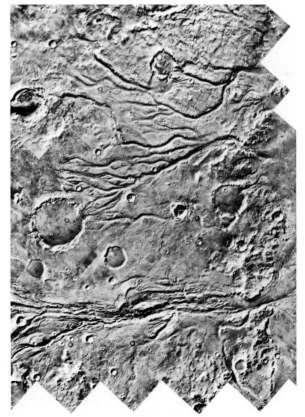

One of the reasons for the choice of the Viking 1 landing site in the Chryse Planitia region shown here was that, from earlier Mariner 9 imaging, it was known that this is a region into which a number of channels had apparently drained in the distant past. The principal Viking goal was the search for evidence of life; an association of life with water and the belief that the channels had been cut by flowing water led to the recognition of the Chryse region as a promising one in which to begin the search. Prior to the landing the region was mapped extensively from orbit to confirm that its characteristics were not inimical to landing safety. That mapping provided this mosaic showing a system of channels rising in the cratered terrain of Lunae Planum and draining eastwards into the Chryse region. The mosaic covers an area of about 100 km by 170 km and is centred at about 55°N, 20°W.

The exotic names attached to the principal channels on Mars are derived from many languages – they are the names in those tongues for the Red Planet. Ma'adim Vallis shown here is located south of the equator near 180°W longitude and, like most of the sinuous channels on Mars, lies near the globe-encircling boundary between the ancient cratered terrain and the younger plains. Its total length is about 200 km and it terminates abruptly well within the cratered terrain. Ma'adim Vallis displays the remarkable sinuosity that characterizes these Martian channels and it is difficult to conceive of mechanisms for its formation that do not involve the flow of a liquid. It is also hard to think of a liquid other than water that might have cut the channels. The origin of the water, if such was the liquid in question, need not have involved precipitation nor atmospheric conditions in which liquid water would have long-term stability. The association of many of the major channels with the cratered terrain-plains boundary, and the generally held idea that the boundary is the result of ground-ice sapping, suggests that the water may have originated from below the surface, released perhaps by magmatic heating. This picture is about 250 km in its horizontal dimension.

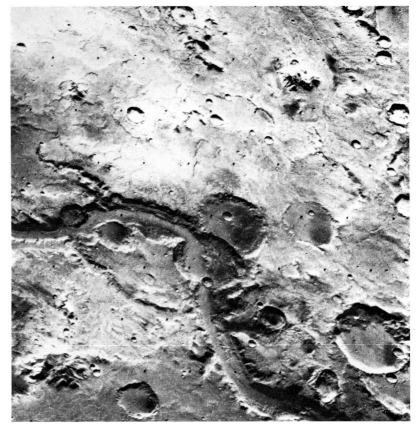

Kasei Vallis is a broad channel to the north of the Viking 1 landing site that flows along the boundary between the ancient cratered highlands and the northern plains. This high-resolution picture, covering an area about 80 km on a side, shows a number of streamlined islands that were presumably shaped by the flowing water which cut the channel. The islands are characterized by long tapering tails and rounded upstream 'prows'. They may be partly depositional in nature, the result of accumulations of material behind an obstacle in the flow.

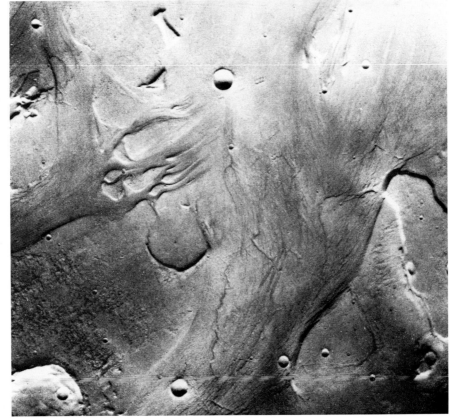

Dunes have been observed in many pictures of Mars taken from orbit. The most extensive field of dunes is that which surrounds the permanent north polar ice cap, covering an area comparable to the Sahara. Elsewhere on Mars dunes appear to be generally confined to the floors of craters such as the ones shown here. The largest crater is named Proctor, located at 48°S, 330°W and about 160 km in diameter. The appearance of Martian dunes is generally similar to common crescent-shaped terrestrial dunes, implying similarity in the mechanism of formation. On both Earth and Mars the sizes of the dunes can vary widely while shapes are generally maintained. Furthermore, within a given dune field on either planet there tends to be considerable uniformity in the patterns that are formed. Given the very thin Martian atmosphere, much higher winds are required to move the fine surface particles that accumulate to form the dunes. The motion occurs as a series of hops in a process called 'saltation'. On Mars the most easily moved particles are calculated to be about one-fifth of a millimetre across (a size that would correspond to very fine sand), requiring wind velocities of 100–300 km/h.

Evidently such high speed winds are not uncommon on Mars.

The photocoverage provided by Viking Orbiter 2 of
the entire northern polar region showed that, unlike
the case in the south, the residual northern polar cap
is surrounded by a ring of sand dunes, lying
between 70°N and 80°N. This colour composite
picture, covering an area of about 135 × 75 km,
shows the margin of this ring of dunes which are
located within the dark band towards the bottom of
the frame. The darkness of the dunes indicates that
the sand-sized material – which is made up of the
more erosion-resistant minerals – is the dark
mineral component of the Martian crust.

Regional dust storms have been recorded in
numerous Mariner 9 and Viking Orbiter images.
Changes in the patterns of light and dark splotches
which cover the Martian surface have been observed
following such storms. This frame is unusual in that
it appears to show rather clearly such behaviour on a
much smaller scale: dust is being picked up from
sites only a few hundred metres across and is being
transported in confined, linear clouds for distances
of many tens of kilometres. It is inferred that bright
surface material is being swept away to reveal the
underlying dark surface, thereby adding to the dark
albedo features that are readily visible in the picture.
The highly irregular shape of the albedo boundaries
apparently reflects their mode of origin. This
picture is about 300 km across. The area in question
is located near the prime meridian, about 900 km
south of the equator.

Mars

The two largest basins on Mars, Hellas and Argyre, are both located in the southern hemisphere. Argyre is the smaller of the two, being about 900 km across. As a result of volcanic flooding the basin is relatively shallow, only about 2 km deep, and smooth. This oblique picture was acquired during the southern winter season and was taken looking across the basin, of which about half can be seen, from the southwest. The basin floor is covered by carbon dioxide frost, probably many centimetres deep. The frost has also accumulated around the bases of the surrounding mountainous region and serves to highlight these features. Paralleling the limb of the planet can be seen an extensive haze, probably composed of water ice crystals, that has formed at an altitude of about 25 km. It is notable that, in general, the atmosphere is very clear and not obscured by the widespread cloudiness that marks the Winter season in the north.

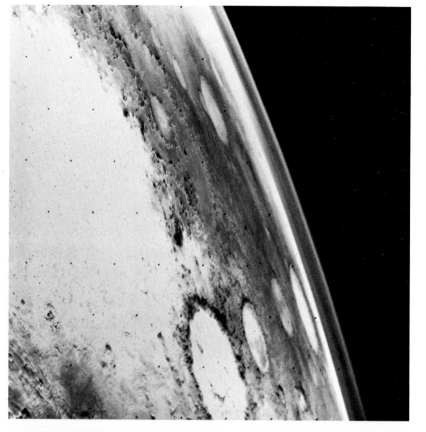

In winter the southern CO_2 polar cap reaches as far north as 40–45°S. The edge of the cap is shown well in this picture of the 250 km diameter crater Copernicus at 48°S, 169°W. The cap is not laid down uniformly, probably as a result of a number of factors: surface albedo, elevation, slope, thermal properties of the soil and local winds.

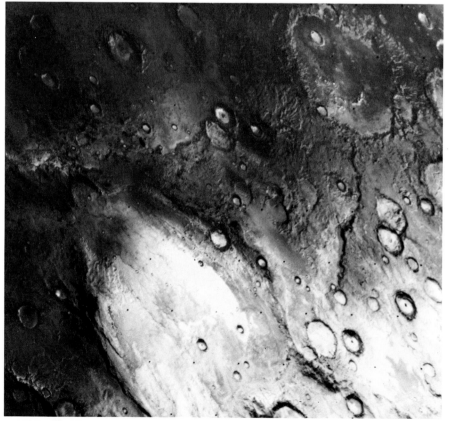

The springtime retreat of the Martian south polar cap can be observed by Earth-based telescopes with sufficient resolution to identify large-scale features. The retreat, which is not a simple circularly-symmetric shrinkage, appears to follow a closely similar routine each year, with a predictable rate of retreat and with the repeated appearance of certain features. As the cap gets smaller it has been observed that a bright island of frost is left outside the cap at the same location (70°S, 330°W) and at the same time (late spring): this region was thought to be of high elevation and was named the Mountains of Mitchel. In this Viking Orbiter mosaic is a close-up view of the region in question at just about the time when the region becomes detached from the main cap. It appears as a bright peninsular and is located near the large crater Main. There is no evidence that the region is elevated and, in fact, the contrary seems more likely. A depression would favour the accumulation of frost, especially as a result of wind-drifting of frost from other areas. A greater depth of frost cover would, naturally, take longer to sublime in the spring.

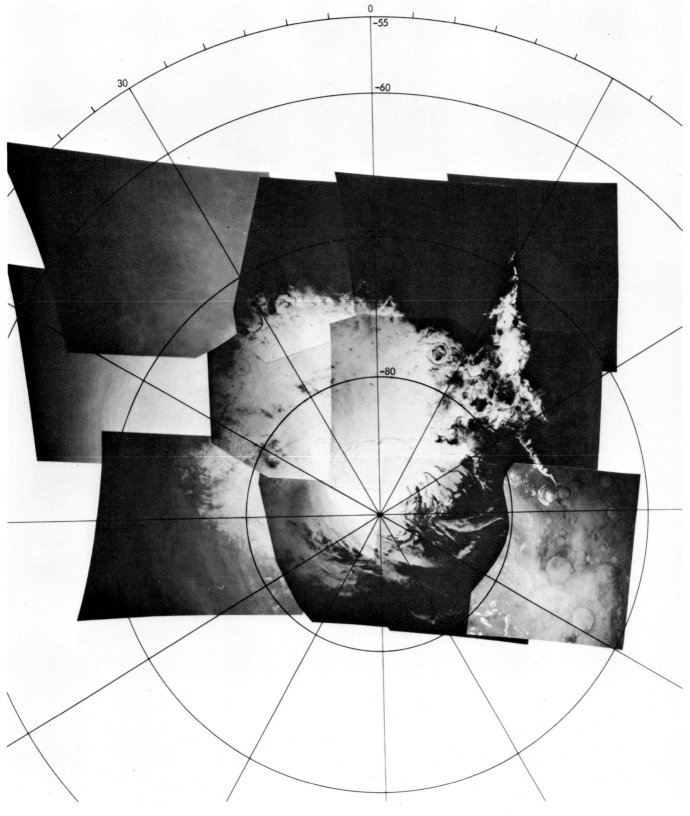

Mars

The layered sedimentary deposits in the north polar region take on the appearance of an abstract composition in the summer months when the frost disappears from the south-facing slopes and when striking patterns are created by frost that is redeposited by the prevailing winds. The layering of the sediments – which might be taken for fine brushwork – further adds to the effect. This particular view is centred at about 79°N, 341°W and covers an area of 63 km × 88 km.

In the long Martian summers neither the northern nor the southern polar cap disappears entirely. Temperature measurements made from orbit indicate that the northern remnant is composed of water ice while, for reasons not so far understood, the southern remnant appears to be largely carbon dioxide. The high inclination and orientation of Viking Orbiter 2's orbit about Mars allowed it to take high-resolution pictures of the northern summer cap; a few colour composite pictures were produced including the one shown here. The terrain near the poles consists of thick sediments deposited in layers and eroded into a complex pattern of scarps and valleys. In summer the Sun sublimes away the ice on equator-facing slopes to leave a contrasting scene of white ice-covered plains and red, denuded slopes. A close-up view (about 65 × 30 km in dimension) of such a region is seen here. The

sinuous boundary between ice and bare ground running across the picture is the top of a 500 m high cliff. In the cliff face can be seen evidence of the layered deposits which average about 50 m in thickness. These deposits are thought to be accumulations of dust, blown poleward from lower latitudes, and ice. The layering is interpreted to imply major periodic changes in atmospheric conditions, possibly the result of periodic variations in both the shape of the Martian orbit about the Sun and in the obliquity of Mars (the inclination of the spin axis).

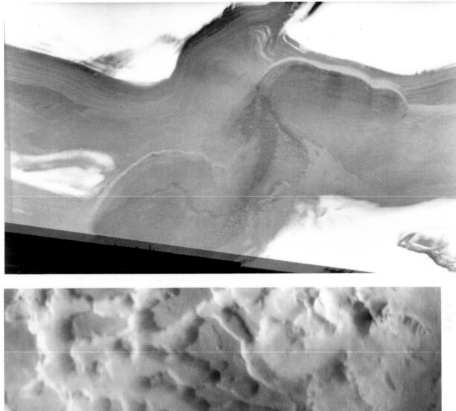

The Viking Orbiters frequently recorded the presence of low-lying condensate cloud around and within the Valles Marineris in the early morning hours. This colour composite picture, taken early in the Viking mission during southern winter, shows the complex of canyons – Noctis Labyrinthus – at the western end of Valles Marineris. The condensate nature of the mist – almost certainly water ice given the temperatures anticipated at that time of day – is evident from its white colour. The mist is mostly down within the canyons but does spill over onto the surrounding plain. Since the Martian atmosphere, though ultra-dry by terrestrial standards, is nevertheless often near saturation, the creation of mist in low-lying regions during the cold Martian nights is not too surprising.

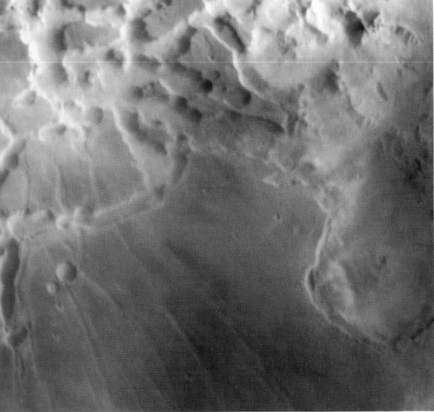

Mars

The Viking Orbiter mosaic shown here was
acquired two days after the initial local dust storm
activity associated with the global dust storm of
February 1977. The mosaic covers a large slice
across the planet running from near the equator
(top), where part of the Vallis Marineris canyon can
be seen, to the far southwest. One or two frost-filled
craters of the retreating southern polar cap can just
be seen at the bottom right of the mosaic. The dust
clouds display considerable structure. Shadows
indicate large-scale vertical relief in the clouds. The
billowing cellular appearance of the clouds to the
south imply convective (overturning) motion.
Although the area to the north has a smoother cloud
cover, it again becomes more structured towards an
apparent storm 'front' south of the great canyon.
Shadows in this area imply cloud heights of
15–30 km.

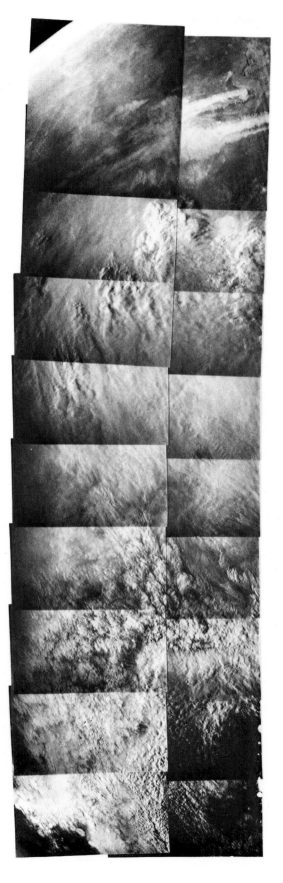

A rare cloud formation is shown in this Viking Orbiter picture taken in early summer near the edge of the northern remnant ice cap. The spiral cloud formation at top left resembles meteorological satellite pictures of terrestrial extra-tropical cyclones. Temperatures measured simultaneously suggest that the cloud is formed of water ice. The radius of curvature of the cloud spiral is about 100 km. This observation, together with one or two others, provides evidence of a wave-type circulation in the summer season. Evidence for such a circulation regime during Martian winter had been provided previously by Mariner 9 images and by pressure measurements made by the Viking landers.

On 20 July 1976, Viking 1 landed on Mars in a region named Chryse Planitia – the Plains of Gold – a geologically complex basin about 2000 km northeast of the Valles Marineris. Relatively smooth when viewed from orbit, the plains contain prominent *mare* ridges, mesas and plateaux. The region also contains 'etched' surfaces, 'knobby' features, low shields of probable volcanic origin, and extensive areas that have been subject to the processes that formed the channels on Mars. Two sets of channels terminate in the area. The top photograph is a view to the northeast from the Viking Lander 1 spacecraft and shows a panorama that includes not only a typical scattering of angular rocks but also a field of sediment drifts that lend the scene unusual beauty. Prominent in the picture is a large rock capped with a layer of dust. *Big Joe*, as the rock came to be known, is about 2 metres across and is located about 9 metres from the lander. Close inspection (bottom) shows that Big Joe is actually two similar rocks side by side. The top image was acquired on 17 February 1977 to document deep trench digging activities that had taken place several days earlier. The trenches are seen to the right of a boom that holds the meteorology sensors for the measurement of temperature, pressure, and wind speed and direction. The colours of the Martian surface are better determined by the landers than by the orbiters since each lander was equipped with a colour test chart. The sky above the horizon always has an orange-brown cast as a result of suspended dust.

158

Mars

The second Viking Lander touched down on Utopia Planitia (48°N, 226°W) on 3 September 1976. The landing site lies within the vast plains region that typifies the northern hemisphere of Mars. Chosen for its apparent smoothness, Utopia is generally lacking in impact craters; the dominant topographic elements of the area are 'hummocks', 'knobs', irregular pits, and domes that may be of volcanic origin. Utopia is clearly not a well-understood region. As can be seen from this view to the south, Viking 2 sits on a flat plain of sediments littered with boulders; these may be ejecta from the large crater Mie, about 200 km to the east, or alternatively may be the remnants of ancient lava flows. (Many of the boulders are deeply pitted like sponges; such pits, or *vesicles*, are commonly formed by the solidification of a frothy, gas-filled lava.)

To the northeast, Viking 2 looks out over a boulder-strewn landscape to a horizon that seems tilted – an artifact of the spacecraft's landing orientation: one leg is evidently resting on a rock creating an 8° tilt. Most notable in this picture is a large boulder in the left middle of the frame. This rock is about 1 metre across and close inspection in higher resolution black and white pictures shows that it is layered. Other rocks in the scene are vesicular in appearance while some have been faceted by the winds to form *ventifacts*: a good example is the small pyramidal rock about 1 metre to the left of the large boulder.

Mars

A clear view of the region of drifts seen in the Viking 1 figure is provided here. Lighting from the rising Sun emphasizes the drifts clearly. Looking closely at the eroded drift faces it can be seen that they are layered, indicating that the sediments were deposited in a series of events, presumably related to changing wind directions. At this time the drifts are not accumulating but, rather, are being eroded, a process that exposes the internal stratification.

A more typical view of the Martian landscape is provided by this picture taken by the Viking Lander 1 spacecraft. Early morning lighting brings out excellent detail in the field of irregular rocks that are scattered across the surface all the way to the horizon. These rocks were probably emplaced as a result of an ancient impact event that excavated local bedrock. Another possibility is that the rocks here were deposited by an enormous flood in the distant past. After emplacement the rocks have been subjected to weathering processes.

The two facsimile (scanning) cameras on each lander are mounted on short masts on one side of the three-sided spacecraft. Views acquired by looking back across the lander are inevitably highly obscured by the complex protuberances on top of the machine. This is a view towards the back of Viking 1 and it shows, framed between the support mast of the communications antenna and the top of one of the lander legs, a coarsely pitted block that is partially buried in fine-grained sediment. It is likely that the sediment is a wind tail formed on the lee side of the rock. The effect of wind is variable: other rocks near the lander, seen in other pictures, are observed to be sitting in erosional depressions.

There is little information in this Viking 2 lander picture of a Martian sunset but it has a mystical quality that makes it irresistible.

Mars has two small moons, named Phobos (fear) and Deimos (panic) after the attendants of the war god. Phobos (top) has a mean diameter of 22 km and orbits Mars every 7 hr 39 min. Deimos (bottom) is smaller (14 km mean diameter) and is in a higher altitude orbit (rotation period 30 hr 17 min). Both are locked into synchronous rotations, like our own Moon, such they keep the same face always toward Mars. The Martian moons appear to be similar objects: each is extremely dark and lacking in colour, and each has a density of less than $2 \, g/cm^3$, suggesting that their composition is akin to carbonaceous chondritic meteorites. Such meteorites are thought to originate from the asteroids and one speculation is that Phobos and Deimos are captured asteroids. There are, however, arguments that militate against this idea. In particular, the orbits of both lie very close to the equatorial plane of Mars and there is no obvious mechanism to create a significant change in the inclination of a captured body's orbit to create this otherwise unlikely situation. The surfaces of Phobos and Deimos are thought to consist of a layer of debris created by innumerable impact events. Although both are saturated with craters, Deimos has a smoother appearance because some of its craters have been partially filled. Phobos exhibits a pattern of striations that are thought to be cracks resulting from the formation of the largest crater on that moon, Stickney.

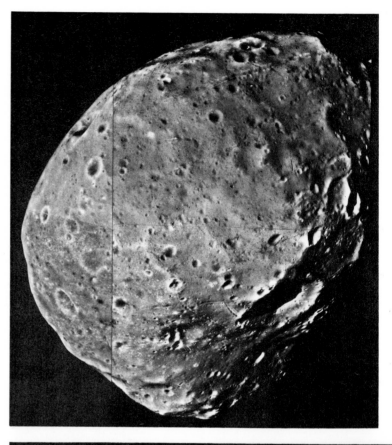

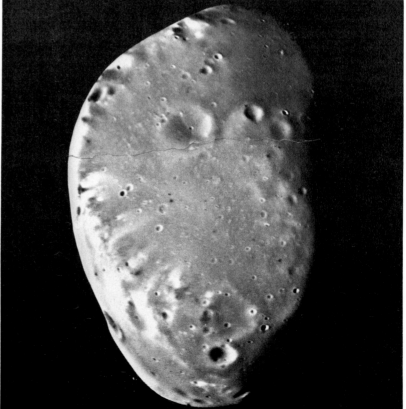

Given the small size of Phobos, Stickney (shown here) is an immensely large crater (10 km in diameter) that adds to the irregularity of an already potato-shaped object. The network of linear grooves that uniquely characterizes the inner Martian satellite appears to radiate from the antipodal point of Stickney; the grooves are evidently fractures formed by the impact event that excavated the crater (and which nearly shattered the entire moon).

In a remarkable example of precision navigation, the Viking Orbiter 2 spacecraft was manoeuvred to fly within 30 km of Deimos in October 1977. The ultra-close flyby was carried out to allow very-high-resolution imaging of Deimos and to allow an accurate determination of the moon's mass (and thus density to provide information about composition). This picture, taken during the encounter, is one of the highest-resolution pictures ever taken of another planetary body by an orbiting spacecraft: the frame is only about 1.2 km × 1.5 km in size and features as small as 3 metres can be resolved. The picture shows a surface saturated in craters, many of which, however, are largely filled with dust to leave only a ghostly outline. Boulders, presumed to be impact ejecta, as large as a house (10–30 metres across) are strewn about the surface.

This shaded relief map of Mars was produced using
Mariner 9 photographs and the radiotracked
position of the spacecraft. Shaded relief has been
portrayed with uniform illumination with the Sun
to the West. No attempt has been made to duplicate
precisely the colour of the Martian surface, although
the colour does approximate to it. 1 degree equals
59 km.

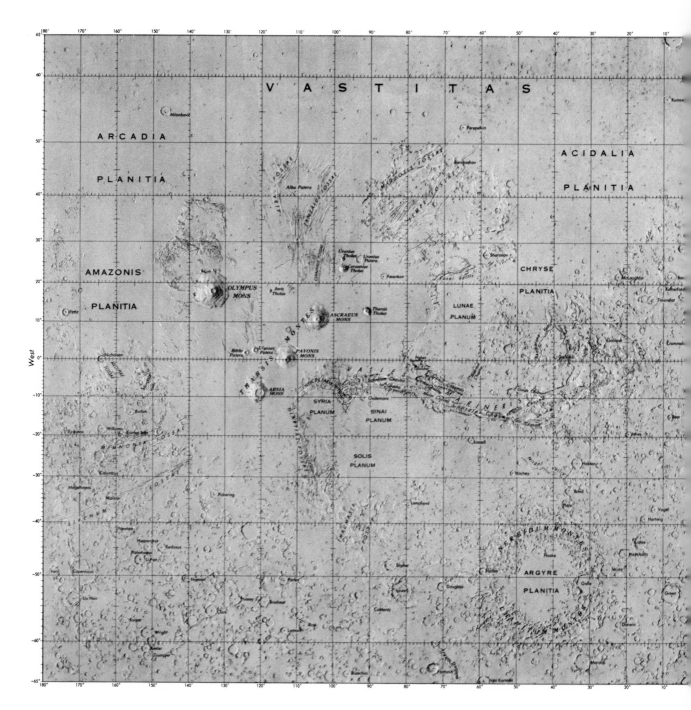

Mars

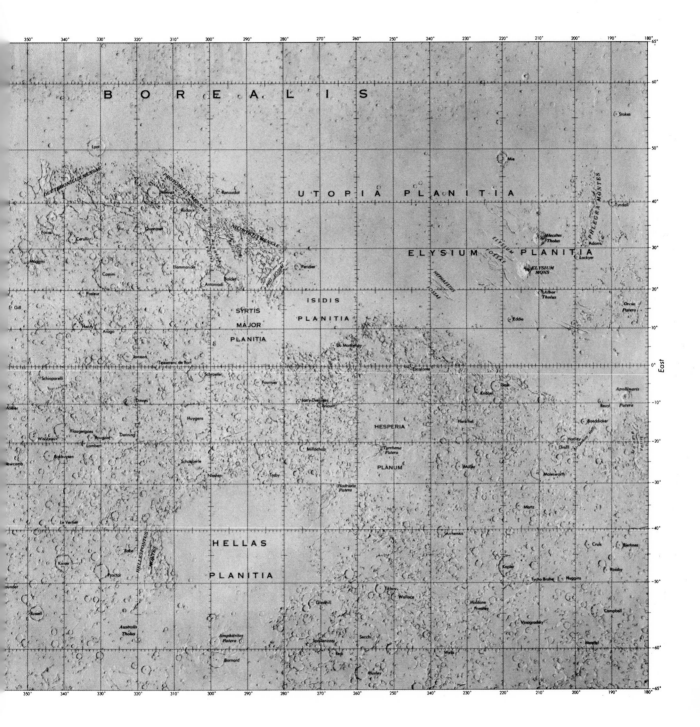

The Jovian system

Jupiter Jupiter is a lambent yellow object in the Earth's night sky. It has been realized since ancient times that this is a very distant planet, approximately fifteen times as far from the Earth as Venus, which alone among the planets significantly exceeds Jupiter's brilliance to the observer on Earth. The fact that Jupiter appears brighter than Mars, when both are at opposition, is principally due to its enormous size. With over a thousand times the volume of the Earth, Jupiter contains more matter than all of the other planets combined. Its mass is an impressive 1.9×10^{27} kg, and it measures 142 800 km across at the equator (317.9 and 11.2 times the corresponding values for the Earth, respectively). These and other statistics on Jupiter are summarized in the following table:

Mean distance from the Sun 5.203 AU	Equatorial radius 71 398 km
Orbital period 11.9 years	Polar radius 66 770 km
Inclination of equator to orbit 3.07°	Mass 1.901×10^{27} kg
	Mean density 1.33 g/cm³
	Period of rotation 9 h 55 min 29 s

Being in a superior orbit with respect to our own planet (that is, further from the Sun), Jupiter always presents a full or nearly full disc to an observer from the Earth. Through even a small telescope, the banded appearance is quite evident, and some hint of the variety of colours which wax and wane between the darker belts is easily obtained. After this initial impression, the next feature to arrest the first-time observer is usually the cluster of brilliant satellites which orbit the planet, the four largest of which were observed by Galileo as long ago as 1610. These are large enough to be considered as planets in their own right, and are of such spectacular appearance and scientific interest that they will be described in detail later.

The face of the planet Jupiter is entirely covered by cloud. In contrast with the bland appearance of Venus, however, the clouds of Jupiter appear full of multi-coloured detail at visible wavelengths. The patterns and colours change constantly on time-scales of hours or days, although always retaining the basic character of the banded pattern. Simple considerations, foremost amongst which is the low mean density, show that the atmosphere must be very deep, comprising at least a large fraction of the radius and perhaps the entire planet. For this reason, Jupiter and its smaller and more distant companions Saturn, Uranus and Neptune, are often called 'gas giants'. The features which we see in pictures of any of these planets are manifestations of the *chemistry* and *meteorology* of the atmosphere, having no direct relationship with the properties of the surface, even if a surface exists at some great depth.

The first order chemistry of the Jovian atmosphere is quite simple: the planet has a composition akin to that of the Sun, made up principally of hydrogen and helium in the proportion of about four to one. Exact solar composition is not expected, however, because it is supposed that the protoplanetary condensation that eventually became Jupiter needed the prior formation of a large nucleus from the dust and ice grains that were available in the outer Solar System. It is thought that a protoplanetary core accreted to contain about ten times the mass of the Earth. This would induce a local gravitational instability, leading to the collapse upon the core of large quantities of the solar nebula gases (having the same composition essentially as the Sun). Thus the overall composition of Jupiter would not be precisely solar in proportions but would have a relative excess of the materials from which the core was made. Calculations suggest that the core of Jupiter contains about a tenth of the planet's mass; hydrogen and helium make up most of the rest of the planet. At great depth within Jupiter pressures are sufficient to break up the hydrogen molecules and even the resultant hydrogen atoms lose their electrons and are left as bare protons – the metallic phase of hydrogen. This is thought to take place near a depth of about 50 000 km.

The colouration of the clouds – predominantly various combinations of white, yellow, brown and reddish tints – speaks of an abundance of minor constituents, some of the simplest of which have been detected by infrared spectroscopy and are shown in the following table:

The composition of the atmosphere of Jupiter as determined by infrared spectroscopy

Molecule	Abundance relative to H_2	Molecule	Abundance relative to H_2
H_2	1	H_2O	1×10^{-6}
HD	2×10^{-5}	C_2H_2	8×10^{-5}
He	0.05–0.15	C_2H_6	4×10^{-4}
CH_4	7×10^{-4}	CO	2×10^{-9}
CH_3D	3×10^{-7}	PH_3	4×10^{-7}
$^{13}CH_4$	6×10^{-6}	GeH_4	6×10^{-10}
NH_3	2×10^{-4}		

The colours require the presence of more complex molecules than those in the above table, although not necessarily in more than trace amounts. Current theories favour sulphur and phosphorus, in elemental and polymer forms, but experimental data are lacking.

Simple (or even complex) amino acids are possible (some say likely) cloud constituents. Photometry, radiometry and spectroscopy from spacecraft have begun to shed light on this and other intriguing possibilities.

The first spacecraft to visit Jupiter was Pioneer 10 in December 1973, followed by its nearly identical sister craft Pioneer 11 in December 1974. The Pioneers were low-cost exploratory missions, carrying fairly simple instruments. Their purpose was principally to show that human ingenuity was up to the task of navigating across the asteroid belt and surviving the hostile Jovian radiation belts. Two instruments are of interest for our present purposes: the imaging photopolarimeter and the infrared radiometer. The former imaged Jupiter in reflected sunlight at visible wavelengths, the latter in Jupiter's own thermal emission in the infrared spectrum. Both built up their pictures a line at a time, using the motion of the spinning spacecraft to obtain each swath. The infrared pictures are of much poorer quality than the visible; this is because infrared technology still lags far behind visible imaging techniques. It shows, nevertheless, the basic banded structure quite clearly. In infrared pictures, the brighter regions are hotter than the darker. Comparing the visible and infrared images, it becomes clear that the light regions (called 'zones') are strips of cloud which appear cold because they are high in the atmosphere, where most of the heat is easily lost to space. The darker, narrower 'belts', are relatively clear regions of thinner or deeper cloud cover. This interpretation is confirmed in spectacular fashion by infrared images taken at the wavelength of five micrometres (about seven times longer than visible light). These have only been obtained from the Earth to date, and so the spatial resolution is not as good as the Pioneer images. Nevertheless, the belts appear spectacularly bright, in some localized regions bright enough to correspond to emitting temperatures of $0°C$ or higher. This is warm indeed on a planet whose average radiometric temperature is close to $-140°C$. Recently, the analysis of this type of image has revealed that the Jovian clouds, categorized by their temperature, are of three types: cool (temperatures in the region of $-130°C$), intermediate (around $-40°C$) and warm (around $20°C$). Furthermore, the three temperature regimes are correlated with the visible colours; cool clouds are predominantly white, intermediate clouds brown, warm clouds so deep in the atmosphere that Rayleigh scattering in the thick mass of atmosphere above them makes them appear bluish (the same effect which colours the sky on Earth). The inference is that there are at least three layers of cloud on Jupiter, at different heights in the atmosphere, and having different compositions. Calculations of the expected chemical equilibrium state of the atmosphere, assuming a 'primordial' or near-solar composition, give a strong clue as to the chemistry of the layers. In simplified terms, the theory says that the common elements like nitrogen, oxygen, carbon and sulphur should be present as their hydrides, e.g. NH_3 (ammonia), H_2O (water), CH_4 (methane) and H_2S (hydrogen sulphide). This follows because hydrogen is present in such abundance that the 'reducing' reactions which form these species are more probable than the 'oxidising' reactions which produce, for example, CO_2 in large quantities on the terrestrial planets. The reactions generally take place most efficiently deep in the atmosphere where temperatures are high, then the products are transported upwards to cooler levels in convection cells driven by the planet's internal heat (see below). The least volatile species condense out at the lowest, warmest altitudes while some, like methane, never get cold enough to condense on Jupiter, even at the top of the convection cells (the tropopause) where the temperature is, at around $-160°C$, the lowest anywhere on or in the planet. Others, like NH_3, H_2O and H_2S, condense at temperatures found near the tropopause or within 100 km or so below it, and form the cloud layers visible from above. Solid ammonia, which is white, forms a cirrus-like cloud at temperatures near $-130°C$; H_2S, as the compound ammonium hydro-sulphide ($NH_3 . H_2S$), which turns brown on exposure to sunlight due to the formation of polymerized sulphur, condenses at $-50°C$; and water, presumably containing dissolved ammonia ($NH_3 . H_2O$) forms clouds at temperatures in the neighbourhood of $0°C$. Non-equilibrium chemistry, involving a multitude of compounds of many elements, may be required to account further for the myriad colours forming the subtle contrasts which are blended into this simple picture.

The Jovian meteorology is no less fascinating than the chemistry, and the two are linked through the transport of reactive materials by the circulation, the temperature dependence of chemical reactions, and the release of latent heat and chemical energy by thermodynamic and chemical processes. The photographs reveal currents and vortices within the basic belt – zone structure, which are part of a vastly deep and restless atmospheric system. A major difference between Jupiter and the Earth is the fact that the Jovian circulation is not driven principally by the Sun. Jupiter is large enough that it still has not cooled since its formation 4.6 billion years ago; its interior releases heat at a steady rate and so heats the atmosphere from below. Infrared measurements from the Earth and from spacecraft show about twice as much heat energy leaving the planet as is deposited by the Sun. On the Earth, of course, these two components nearly balance. The result on Jupiter, which we see in the photographs, is the formation of tremendous convection cells similar in concept to those in a pan of nearly boiling water. In the rising regions, warm moist air laden with H_2O, H_2S and NH_3 rises through the cooler levels where these volatiles condense and then descends back towards the interior. The descending air is dry (depleted of volatiles) and also, because it is travelling downwards, tends to sweep any cloud particles with it to levels where they can evaporate and disappear. Thus, the descending branches of the convection cells are relatively cloud-free and it is these which we see as the dark belts. The whole display is swirled into bands by the rapid spinning of the planet. Since Jupiter rotates every ten hours, the cloud tops at the equator travel around the rotational axis at an impressive 48 000 kilometres per hour. The Coriolis forces associated with such rapid rotation strongly constrain horizontal motions towards the zonal direction (parallel to the equator) and lead to axisymmetric convection cells. The characteristic horizontal spacing of the belts and zones is a result of stability criteria which depend on ponderable but unmeasured quantities such as the vertical depth of the convection cells. Mathematical models have been devised which reproduce the appearance of Jupiter approximately, but they contain many assumptions and the general circulation of Jupiter must still be considered only marginally understood.

The detailed meteorology of Jupiter could not be studied with the relatively low-resolution images from the spinning Pioneer spacecraft and further progress awaited the pictures from the more sophisticated cameras on board the stabilized Voyagers, which arrived at Jupiter in March and July 1979. (Planetary missions often consist of pairs of identical spacecraft launched close together. Voyager was probably the last to follow this strategy, because spacecraft have evolved to be both more reliable and more expensive.) Voyager pictures of such prominent and apparently permanent features as the Great Red Spot were eagerly awaited and did not disappoint. First observed by Robert Hooke in 1664, this massive cloud bank has persisted ever since, in the same general location. Long discourses have been written on its origin and sustaining mechanism. Early speculators favoured the 'obstacle' theory whereby the spot formed continuously above some vast (and hypothetical) Jovian mountain, or above a floating obstacle such as a fallen asteroid or comet. Current theorists prefer to understand the Great Red Spot as a storm or as a 'solitary' wave motion of long but not infinite duration. In fact, the observational evidence is insufficient for a firm conclusion. In the Pioneer and Voyager photographs, clear evidence is seen for vorticity within the Spot. Its temperature is low, lower even than the 'cool' zone clouds which enclose it, implying that the cap of the phenomenon is higher than the surrounding clouds. Interestingly, it is not, as was once believed, unique. The Pioneers and

Voyagers both photographed several smaller red spots at various locations on the planet.

Other permanent and semi-permanent features of the Jovian cloudscape include the smaller red and brown spots, the *white ovals* and the *white anvils*, the *equatorial jet*, and the more rapid but less visually striking *north temperate belt jet*. The jets are regions where the cloud belts are observed to be rotating in the zonal direction more rapidly than their neighbours. By tracking the motions of features within the jets from the Earth, it has been found that wind velocities of at least 140 m/s occur on a fairly regular basis and may persist permanently. Several other currents of lesser magnitude are also present, as are wave motions on various scales, and can be located in the high-resolution Voyager photographs.

Velocities in the jets fluctuate, typically by around 20%, when a large sample of measurements spanning many decades is considered, but in general the features are always present. The mechanism driving them is not understood in detail, as indeed it is not for the jets in the Earth's atmosphere at similar pressures (of the order of $\frac{1}{10}$ of an atmosphere). Detailed studies of the Voyager images have shown the presence of a great deal of eddy motion (which reveals itself as relatively small-scale vortices in the clouds or as fluctuations in the mean flow) associated with the jets. Their superficial appearance suggests instabilities breaking away from the jets and taking some of their energy. In fact, studies of individual vortices show them rotating faster than the zonal flow and tending to accelerate it. Apparently, eddies are the means by which angular momentum is brought upwards from the deep atmosphere and this process may be most efficient at low latitudes where convective velocities are expected to be greatest, accounting for the equatorial jet.

The white ovals and red spots are themselves eddies of great size and duration. There is evidence for a lifetime in excess of half a century for some white ovals. Most of them rotate anticlockwise in the southern hemisphere and clockwise in the northern; this means that they are high-pressure features like terrestrial anticyclones. The reason for their long life is likely to be connected with the relative absence of diurnal change on Jupiter (because of the short day–night cycle and the distance of the Sun), which has a dissipative effect, and the constant supply of fresh energy and angular momentum from smaller eddies convected upwards. A wide range of more detailed models has been suggested but at present the data to discriminate between them are lacking.

The principal difference between red and white spots may be the depth to which they extend; more energetic disturbances may mix the atmosphere vertically so that the chemical reactions in the clouds never reach equilibrium with the surrounding temperature, resulting in a different mix of compounds and hence a different colouration. Elemental phosphorus, produced in the warm lower atmosphere and 'frozen' into the cold upper clouds before it has time to convert into phosphine (PH_3), is a possible candidate for the colouring matter in the red clouds. Of course, there are a great number of very complex molecules, many of which are colourful, and the presence on Jupiter of some of them cannot be excluded. Indeed, we probably should expect it, especially if the convective cells are deep so that reactions involving high temperatures and a multiplicity of elements and compounds are involved in the cloud chemistry.

The night side of Jupiter is permanently turned away from the Earth and so, even if Earth-based observers had sufficiently sensitive instruments, they have never had the opportunity to search for aurorae and lightning, phenomena which show up brightly in our own night sky. So it was Voyager's cameras which recorded the first evidence for their Jovian equivalents. While it is not surprising to find both phenomena present on Jupiter, it is exciting to have data on them to investigate their strength, distribution and variability in this alien environment to test our understanding of how such behaviour is produced.

The moons of Jupiter

As is appropriate for the largest planet in the Solar System, Jupiter is accompanied by many moons, of which 16 have been discovered to date. There is also a tenuous ring, first seen in images returned by Voyager. Most of the moons are objects only a few tens of kilometers across, objects that seem more likely to have been captured than to have been formed in orbit around Jupiter; if so then their study will not be helpful in better understanding the planet itself. However, the four largest moons – Io, Europa, Ganymede and Callisto – are all believed to have accreted as part of the process by which Jupiter itself formed and, therefore, contain clues about Jupiter's origin and evolution. They are all, moreover, extraordinarily fascinating objects in their own right. Discovered by Galileo (and, as a family, named after him) in 1610 these satellites are in the same size range as the Moon and Mercury, and they occupy near-circular orbits in the Jovian equatorial plane. By contrast, the other moons are typically in highly-inclined, eccentric orbits at great distances from Jupiter. Some are in retrograde orbits and this randomness implies an origin elsewhere in the Solar System, supporting the suggestion that they are captured asteroids.

Two of the smaller moons do have added interest: Amalthea and a Voyager discovery known only as 1979J1. Amalthea was discovered by Barnard as recently as 1892 and lies inward of the innermost Galilean satellite, Io. Although it was not approached very closely by either spacecraft, Amalthea is seen in Voyager images to be an extremely irregular object having dimensions of about $270 \times 165 \times 150$ km. Its surface is heavily scarred by impact events, is very dark, and of a distinctly reddish colouration. Some of the craters are enormous when considered relative to the size of the moon itself; one is 90 km across, another is 75 km wide. The red surface is conjectured to be the result of a dusting of sulphur (originating at Io as will be discussed later) and laboratory simulations tend to confirm this idea. If bulk density information were available for Amalthea it might be possible to distinguish between hypotheses about the moon's origin, namely, capture or *in situ* formation. Unfortunately, the Voyager flybys were too distant to make a determination of Amalthea's mass.

Little is known about 1979J1 which was observed only as a point of light (the radius of the moon is estimated to be 10–15 km) in Voyager images. It is this moon's *orbit* that makes it of especial interest; moving about Jupiter at less than one Jovian radius (R_J) from the cloud tops, 1979J1 is located just outside the outer edge of Jupiter's ring and is thought to provide the physical barrier constraining the ring dimensions. The ring, which is almost uniform in its structure in contrast to Saturn's rings which are made up of countless distinct ringlets, is probably composed of dust grains that are the result of micrometeorite bombardment of one or more tiny moonlets orbiting within the ring. The outer edge of the ring is at 1.81 R_J from the centre of Jupiter and the images also show an inner edge at about 1.72 R_J (6000 km inward), inside of which a faint sheet of material is seen to stretch all the way to the atmosphere of Jupiter. The sheet is evidently made up of dust that is gradually spiraling into the planet. It is the relatively brief lifetime of the ring material that implies that there must be sources of ring material within the ring. One possible source is a small moon (estimated radius 20 km) discovered by Voyager – 1979J3 – which lies within the ring in a region that exhibits an enhanced brightness. The sharpness of the outer ring edge is thought to be the result of the gravitational interaction of the ring particles with the moon 1979J1: ring particles nearing the moon have their orbits distorted so that they subsequently undergo collisions with other ring particles. The overall effect is that the orbits adjust themselves to prevent the further outward migration of particles.

The principal interest in the Jovian moons lies in the four Galilean satellites which will now be discussed. Together with Jupiter itself these moons make up a Solar System in miniature, an analogy that has become increasingly meaningful as we better understand the evolutionary history of Jupiter and the systematic differences among the Galilean moons. As can be seen in the table below, there is a definite trend in the densities of the Galilean moons as a function of their distance from Jupiter, a trend paralleling that for the Sun and planets.

The Galilean moons

	Radius (km)	Mass (gm)	Density (gm cm^{-3})	Distance from Jupiter (km)	Orbital Period (days)
Io	1816	8.92×10^{25}	3.55	412 600	1.769
Europa	1563	4.87×10^{25}	3.04	670 900	3.551
Ganymede	2638	14.90×10^{25}	1.93	1 070 000	7.155
Callisto	2410	10.64×10^{25}	1.81	1 880 000	16.689

The orbital and rotational periods are identical so that each has a hemisphere that permanently faces Jupiter. These characteristics were known, albeit less precisely, before the Voyager flybys and the general conclusions that had been reached about the origin of the Galilean moons remain essentially unchanged in the light of the results of that mission. Specifically, the two inner moons, Io and Europa, are evidently composed primarily of rocky material, lacking any major component of iron. Europa must also have a small admixture of a low-density material: water ice is the most obvious and likely candidate. The observed reflectance spectrum of Europa provides confirmatory evidence. The two outer moons, Ganymede and Callisto, appear to be made up of roughly equal proportions of rock and ice. It is supposed that the increasing ice component with increasing distance from Jupiter reflects a decrease in temperature with distance from proto-Jupiter during the period, presumed to be about 4.5 billion years ago, when the moons were forming in their orbits. Such a temperature gradient fits well with current ideas about the early history of Jupiter.

The Galilean moons are all sufficiently different from one another that a separate discussion is needed for each, starting with the simplest, Callisto, and moving inward to the most bizarre, Io.

Callisto

Callisto is an airless body, about the same size as Mercury, where surface temperatures never rise much above $-130°C$ and whose evolution came to a standstill billions of years ago. It was first seen in detail by the Voyager spacecraft in 1979 and images from these spacecraft are the principal source of information about this remote world.

As yet we have no direct knowledge about the interior of Callisto but we can speculate on this matter with some degree of confidence. Doubtless Callisto would have incorporated its share of radioactive elements when it accreted and calculations indicate that the heat produced by radioactive decay would be sufficient to differentiate the silicaceous and icy materials that are thought to be the major constituents of this moon. Therefore, we expect that Callisto has a large rocky core surrounded by a mantle of water and ice. The crust of Callisto is evidently not pure water ice: it is darker than that of the other Galilean moons, but still twice as bright as our own Moon. Presumably the crust is a mixture of water ice and rocky material. In places meteorite impacts have punctured the crust liberating water which has been spread over the surface to form rays around bright craters. Impact craters, large and small, are in fact the only kind of surface features that have been observed on Callisto. Their abundance suggests that they have been preserved from the earliest era of Solar System history, more than four billion years ago. The appearance of the craters reveal definite differences from craters on the surfaces of the Moon, Mars and Mercury; they are typically much flatter and lack the central peaks that characterize larger craters on the inner planets. Often they have central depressions. There are no basins on Callisto, though we do see evidence of ancient basin-forming events. These differences are attributed to the icy nature of the impacted crust: even at the very low temperatures encountered on Callisto, craters tend to 'relax' and gradually erase themselves.

Nowhere on Callisto are there found smooth plains regions and, as such, Callisto is a unique object among the solid bodies seen to date. Evidently no internal activity has ever achieved sufficient intensity to sunder the crust and reshape the surface. The most striking surficial features on Callisto are a number of bright concentric ring systems, like ripples on a pond, that are evidently the Galilean equivalent of the giant basins found on the Moon, Mars and Mercury. The largest of these, given the name *Valhalla*, is about 3000 km in overall diameter as measured by the distance to which the concentric ridges extend. The actual impact that formed Valhalla was evidently much smaller; it is thought that the original crater was about 300 km across. There is no longer any such crater but only a flat central region of generally higher reflectivity than average. These 'ghost' craters have been given the name *palimpsests*; they were probably formed while the crust was still relatively warm and lacking in the strength need to support significant variations in vertical relief.

Callisto, by its very simplicity provides us with a valuable reference against which to measure the nature of the other solid icy bodies that reside in the depths of the Solar System. Such comparisons start with its neighbour, Ganymede.

Ganymede

Ganymede is slightly larger than Callisto and is, in fact, the largest moon in the Solar System. No atmosphere has been detected. Its composition is presumed to be similar to that of Callisto and a comparable evolutionary path might have been predicted. The Voyager imaging results show that this is not the case; Ganymede has had a much more complex history. Internally, however, we expect the same general situation as for Callisto – a rocky core and an ice/water mantle. The crust is also presumed to be a mixture of ice and rock. Ganymede bears many impact scars but, unlike Callisto, its surface is also characterized by global-scale light and dark mottling. The dark regions are polygonal in outline and similar in many respects to the surface of Callisto. The high-resolution Voyager images show that the dark regions have a heavy crater population implying an ancient origin. It seems likely that the dark regions are the remains of the original crust of this body. They are not analogues to the dark maria of the Moon, which we know to be lava-flooded impact basins, although they resemble them when viewed distantly.

The Voyager imaging coverage of Ganymede is better than for Callisto and perhaps it is for this reason that a wider range of characteristics is observed for Ganymede's cratered terrain: well developed ejecta blankets (some with lobate flows as on Mars) are seen in some cases and sharp rimmed craters with rays are also observed. Both central pits and peaks are found within craters. The imaging resolution also allows the identification of secondary craters in the vicinity of young bowl-shaped craters. At the other end of the age spectrum are the palimpsests, found on Ganymede as on Callisto, although none are in the same size range as Valhalla.

The dark regions on Ganymede bear a clear family resemblance to the surface of Callisto. The brighter units, which make sharp contacts with the dark terrain, are quite different. They are made up of grooved terrain – complicated mosaics of parallel ridges and troughs – with a significant population of impact craters. These features make complex patterns as they curve, branch, and intersect. Though evidently quite ancient, the crater densities and the stratigraphic relationships indicate that the bright grooved terrains are younger than the dark cratered areas. The ridges and troughs have a vertical relief estimated to be a few hundreds of metres and they run for up to a thousand kilometres across the surface. Typically they are kilometres to tens of kilometres in width. The ridges have sharp crests and the wide flat floors of the troughs are hummocky. In some places there is clear evidence that the crust has been sheared to create lateral displacements of as much as a hundred kilometres.

The grooved terrains on Ganymede appear to have been created at the expense of the more ancient cratered crust. This replacement might have occurred either as a result of *in situ* processes – depositional (e.g. volcanism), erosional and vertical crust motions – or, possibly, by a process akin to the plate tectonics that dominate terrestrial geology. Although the evidence is incomplete, some geological analyses of the Voyager imaging data suggest that *in situ* processes were dominant, in particular intense vertical faulting and local shearing, the result of convective upwelling in the mantle at a time when the crust was still thin and weak. There is uncertainty as to whether plate tectonics were involved as the evidence for subduction is controversial. Localized crustal spreading, however, does appear to have taken place with segments of the crust separating and shearing.

The different evolutionary paths of Ganymede and Callisto present us with a major puzzle in view of their basic similarity, bringing to mind another pair of twin planets with divergent histories – the Earth and Venus.

Europa
Seen only distantly by Voyager 1 as a bright globe criss-crossed by dark curvilinear markings, Europa is still an enigma even after the much closer flyby of Voyager 2. This moon is somewhat smaller than its outer companions, being similar in size to our own Moon. The refinement in bulk density determination achieved by the Voyagers has enabled better modelling of the structure of Europa than could be achieved using telescopic data: an ice/water crust of between 75 and 100 km thickness surrounding a rocky interior fits the new data.

The icy surface of Europa bears no resemblance to those of Ganymede and Callisto, lacking entirely the battered appearance of those bodies and having a much higher albedo that implies a crust composed almost entirely of water ice. Besides having an almost complete absence of craters down to the resolution of the Voyager images (about 5 km), Europa appears to lack almost any vertical relief at all; the highly visible global curvilinear markings, when seen under oblique lighting conditions that emphasise vertical structure, show elevation differences of only a few hundred metres. The striping of the surface of Europa is nevertheless very striking in its extent and in its tangled complexity. The individual dark bands are typically between 20 and 40 km in width and they stretch for thousands of kilometres across Europa. A possible origin of the bands may lie in an episode of global expansion where the crust fractured under tension. The fractures might then have filled with water which froze. The degree of expansion that would be implied by the area of the stripes is around 10%. So far no satisfactory explanation for the cause of such an episode has been forthcoming, nor can the evolutionary history of the moon be reconstructed with confidence. The paucity of impact craters is indicative of resurfacing of the satellite by effusions of water which subsequently froze, but the source of such water is a matter of debate. Certainly Europa contains heat-producing radioactive elements, but calculations suggest that normally expected concentrations of these elements would not provide sufficient heat late in the satellite's history to melt an icy mantle or drive a water volcanic cycle. There is, however, another source of heat. The orbit of Europa, like Io, is subject to a forced eccentricity (see the discussion below for Io) which leads to tidal flexing of the satellite as it orbits Jupiter. Studies indicate that this flexing, while not nearly so great as Io experiences, is sufficient to generate some frictional heat. That heat, together with radiogenic heat, is presumably responsible for the water effusions, but whether that water is derived from an icy mantle, a liquid water mantle, or a hydrated silicate core is unknown.

Io

Experience in exploring the inner Solar System has confirmed our expectation that the degree of geological evolution of planetary bodies is proportional to size – small planets like the Moon and Mercury reach a senescent state early as a result of rapid internal cooling whereas larger bodies, like the Earth (and, most likely, Venus), continue to evolve to this day. In the outer Solar System this generalization still has some validity but, there, we find that bodies as small as the Moon and Mercury can evolve to considerable complexity: Jupiter's Ganymede and Saturn's Titan (see next chapter) are proof of this. Io is an extreme example.

Intimations that Io might turn out to be a most unusual object had been forthcoming since the mid 1960s when it was recognised that Jupiter's decametric radio emissions (discovered a decade earlier and associated with the Jovian magnetosphere rather than with the planet itself) are modulated in intensity according to the position of Io in its orbit. Theoretical studies were undertaken of the nature of the interaction of the Galilean moons with the magnetosphere in which they are buried (the magnetic field of Jupiter sweeps magnetospheric plasma past the moons as the planet rotates) and it was recognised that the moons could act as generators of electrical current, each driving a current system through the conducting path made up of the moon, the Jovian magnetosphere and the Jovian ionosphere. In the case of Io the current was estimated to amount to a million amperes. Another peculiarity was the reported post-eclipse brightenings of Io – short-lived increases in visual brightness following emergence of the moon from Jupiter's shadow. And, in 1973, spectroscopic measurements using a ground-based telescope led to the discovery of a halo of sodium atoms surrounding Io. The sodium cloud, which emits a yellow glow like a street lamp, was found to form a partial Jupiter-encircling torus centred on Io. Emissions from potassium and from ionized sulphur were also detected soon after.

The measurements made by the dozen instruments on the Voyager spacecraft have shed a good deal of light on all of these phenomena (with the exception of the post-eclipse brightening reports) and have confirmed the existence of a mega-ampere current near Io. It is, however, through the imaging observations that we have learned directly the extraordinary nature of the moon: Io is a body that is convulsed with continuous volcanic activity. Resembling a giant neapolitan pizza, Io's surface is an endless volcanic plain, painted in shades of red, brown, yellow and white, and randomly scattered with dark-floored calderas, vents, lava flow sheets and lava lakes.

The evidence for the *current* volcanic activity is straightforward and indisputable: volcanic eruptions are recorded in many images – plumes up to 280 km high on the planetary limb, umbrella-shaped fountains in oblique views, and diffuse brightenings in vertical views. Eight eruptions were observed by Voyager 1 to be taking place simultaneously and, several months later, Voyager 2 determined that six of the eight were still active. Concurrent observations by the Voyager infrared spectrometer identified local 'hot-spots' on the Ionian surface near volcanic features. It is even possible now to monitor the volcanic activity using ground-based telescopes.

The explanation for this quite fantastic degree of volcanic activity is related, not to any extraordinary abundance of radioactive elements in Io's interior, but rather to the nature of the orbits of the Galilean moons. The orbital periods of Io, Europa and Ganymede have evolved to a state of synchronism whereby Io orbits Jupiter in half of Europa's period which, in turn, is half of Ganymede's period. As a result of the gravitational interaction between Io and the other two moons (chiefly Europa) Io is forced into a slightly eccentric orbit. If Io were in a perfectly circular orbit, it would keep the same hemisphere always pointing toward Jupiter and the huge tidal bulge induced by Jupiter's gravity field would remain fixed. The eccentricity of the orbit causes the amplitude of the bulge to vary so that the interior of Io experiences a constant flexing which, as a result of friction, generates heat in amounts that are estimated to be sufficient to melt most of the interior. Interestingly enough, the theoretical analysis of this phenomenon was published just a few days before the Voyager 1 flyby of Io and the volcanism was, in fact, predicted. It seems likely that the tidal heating of Io has taken place for much of the lifetime of the body.

The eruptions seen on Io appear to consist of the continuous explosive ejection of gas and fine particles at very high velocities into the low gravity and near vacuum of the Io environment. Ejection velocities of about 1 km per second have been estimated. The gas and dust return to the surface in a fine spray that, according to some estimates, resurfaces Io at a rate of about 1 cm every 3000 years. Eruptions have also been observed on the flanks of Io's most prominent volcano, *Pele*. The most important process, however, leading to the resurfacing of Io may be direct inundation by lava flows, of which there are numerous examples in the images. Unquestionably, the Ionian surface is extremely youthful so that there is no evidence of the original state of the moon. It seems quite likely that the entire body has been volcanically processed (i.e. erupted on the surface and subsequently reburied and returned to the interior) during the last 4.5 billion years leading not only to the loss of the original form of the moon, but also to the loss to space of most of the volatile substances normally associated with planetary bodies – water, nitrogen, carbon dioxide, and neon in particular.

The explosive volcanic plumes demonstrate that some volatile species remain; the most likely candidates are sulphur and sulphur compounds. Sulphur is a cosmically abundant element but one that does not play a prominent role in terrestrial volcanism because, it is believed, most of the Earth's sulphur is in its core. On Io there are several lines of evidence that argue strongly for the role of sulphur: ionised sulphur and oxygen have been identified in the torus at Io's orbit, sulphur dioxide in its gaseous form has also been discovered, and there is evidence for the presence of sulphur dioxide frost on the surface of Io. Also, the range of hues recorded in Voyager images of Io's surface is consistent with the range that can be assumed by the different allotropes of elemental sulphur (the red colouration of Amalthea may also originate from sulphur lost from Io). So, there seems to be little doubt about involvement of sulphur. There are, however, dis-agreements about the form that volcanic processes on Io take. One suggestion is that the crust of Io is composed of solid sulphur and sulphur dioxide floating on a kilometres-thick layer of molten sulphur that, in turn, overlies a silicate subcrust. In this model the volcanism is thought to involve primarily the sulphur compounds. Another model invokes silicate as well as sulphur volcanism; it is proposed that the crust is made up of interbedded silicates and sulphur with flows of both on the surface. Certainly some of the imaging observations, which show vertical relief on the surface of as much as two kilometres, appear inconsistent with the mechanical properties of a crust composed of pure sulphur and overlying a warm interior. Most likely no single simple model will prove to be sufficient to account for this unique Solar System object.

The giant planet Jupiter, photographed through the 24-inch telescope on Table Mountain, California. Except for the rather unrealistic colour balance – a common problem for Earth-based photographers of planets – this picture conveys how Jupiter appears to the naked-eye observer using a large telescope under close to ideal conditions.

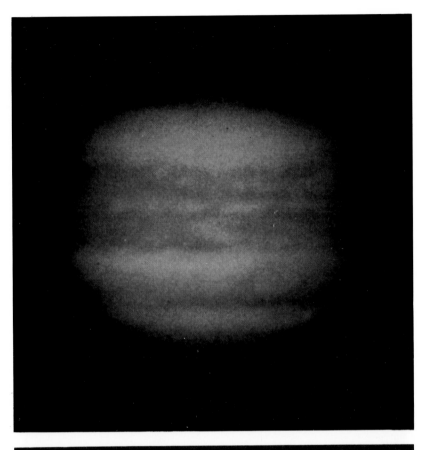

A composite of several black and white exposures of Jupiter, carefully superimposed to reduce the blurring effects of the Earth's atmosphere. Pictures like this have long served to confirm the naked-eye observer's impression that the features visible on Jupiter are due to multiple layers of cloud. No surface is ever seen; it is likely that Jupiter doesn't have one in the usual sense. Instead the density increases gradually to very high values at the interior.

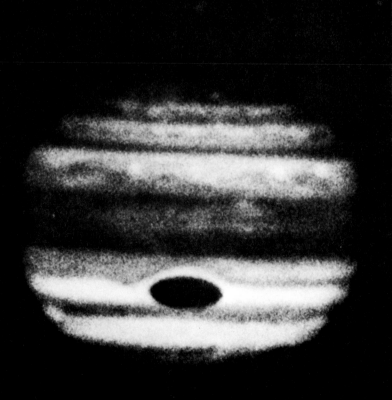

The Jovian system

A picture of Jupiter obtained by Pioneer 10 from a distance of 1.2 million miles. The colour is partly synthetic – Pioneer imaged the planet in only two primary colours (red and blue). The third (green) was added artificially until the mix produced the same general colouration which is seen from the Earth. In this view, the Great Red Spot is just passing into darkness. This is the best-known permanent feature on Jupiter after the belts and zones – it was first observed over three hundred years ago, and may be much older than that. Its origin has been controversial for centuries.

A false-colour picture (left) built up from Earth-based measurements of infrared radiation from Jupiter, in this case at a wavelength of 5 microns (visible light is about 0.4 to 0.8 microns). As in the infrared pictures of Venus we are looking, not at sunlight, but at the heat radiation actually emitted by the planet. Progressively brighter (therefore hotter) areas are represented by colours ranging from black through red to yellow. As the accompanying photograph (right) shows it turns out that the contrasts are the opposite of those in the pictures taken at visible wavelengths, with belts brighter than zones. The reason is that the

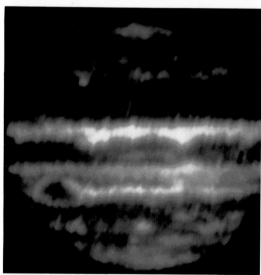

atmosphere gets warmer with increasing depth, so the high, white ammonia clouds emit less infrared energy than the deeper, reddish-brown clouds. Jupiter shows enormous contrasts at this wavelength, equivalent to hundreds of degrees centigrade of temperature difference. The reason is that the gases in the Jovian atmosphere are highly transparent at this wavelength, so the telescope looks deep down through gaps in the clouds to levels on Jupiter where the temperatures are similar to those on Earth.

Jupiter revealed with unprecedented detail by the cameras of Voyager 1 as the spacecraft entered the domain of the giant planet in early 1979. This picture was taken on February 1 from a distance of 20 million miles (about one-twentieth of the distance from Jupiter to Earth). The colours are shown fairly realistically; white and salmon pink are common, with some regions quite red and others yellowish. Regions where the highest cloud layer is broken (as revealed by infrared measurements) appear grey or bluish. The possible causes of the colours are discussed in the text.

185

Voyager 2 picture of the Great Red Spot, showing detail within the spot itself and also some nearby 'white ovals'. The Spot is thought to be a storm, or similar atmospheric disturbance, of very long duration. Its red colour may be due to the presence of phosphorus, mixed up from deeper in the atmosphere by turbulence associated with the storm. Note also the white ovals, and the multi-coloured streamers of clouds. When several pictures similar to this are viewed in rapid succession, the resulting 'movie' shows the spiral structures inside the Spot and the ovals rotating while the streamers 'billow'.

A region to the southeast of the Great Red Spot, shown with the colour contrast stretched (exaggerated) to reveal details of the atmospheric structure. Apart from the variety of cloud types and colours, the most striking aspect is the apparent tendency of some features to pass through each other without mixing, more like immiscible liquids than Earth-type clouds.

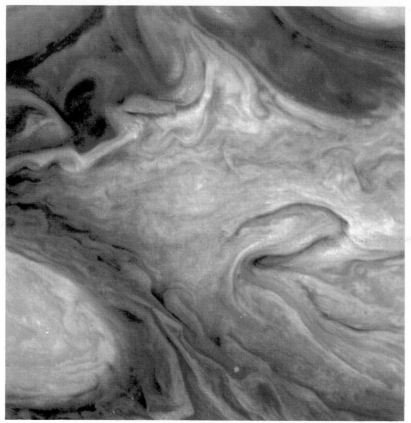

Another Voyager close-up of the Great Red Spot region. Wave motions on several different horizontal scales can be discerned from the cloud patterns. The smallest details are less than 200 km across.

The Jovian system

A red spot in the northern hemisphere, observed by Pioneer 10. Except for its lesser size and weaker colouration, this and other red spots on Jupiter show a strong resemblance to the Great Red Spot, suggesting a common origin.

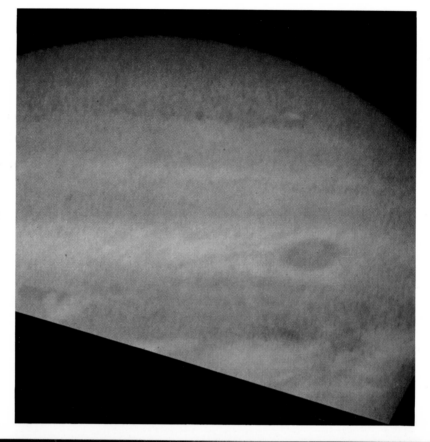

Cylindrical projections, i.e. with linear, perpendicular latitude and longitude scales, showing Jupiter in February (top) and June (bottom) 1979, as observed by Voyagers 1 and 2 respectively. Each picture is built up as a mosaic of five images obtained approximately 2 hours apart during a 10 hour rotation of the planet. The longitude scales are aligned, so that the relative motion of various features in different latitude bands can be seen.

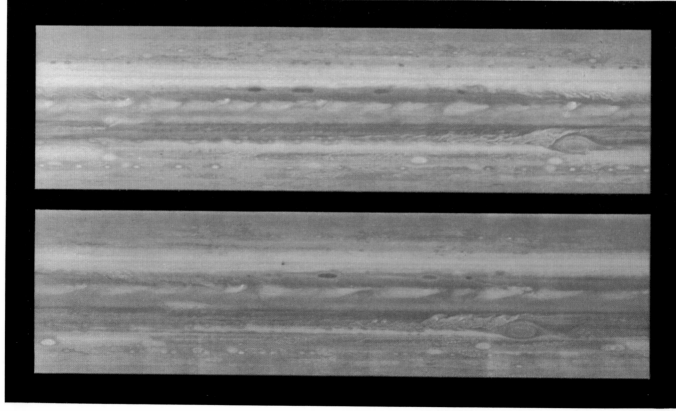

Night-time on Jupiter – with massive versions of two well-known terrestrial phenomena, aurora and lightning, showing up clearly. The north pole is near the centre of the bright arc, which corresponds to aurora borealis much more intense than our familiar northern lights. Similarly, the lightning bolts are much brighter than average earthly flashes. The mottled appearance of the image is caused by electronic 'noise' in the camera system. This is always present in images, but is only visible when excessive contrast stretching is required to bring out features in very dimly illuminated scenes, as here.

The ring of Jupiter, discovered by Voyager 1 in March 1979, is seen here in a Voyager 2 picture designed to view it under good conditions of illumination. Like those around Saturn, but much more tenuous, the Jovian ring is about 6500 km wide and probably less than 10 km thick. The bottom picture shows the scale of the ring, and the portion photographed.

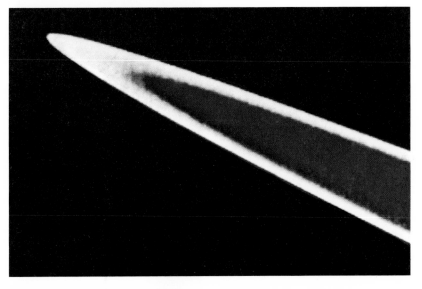

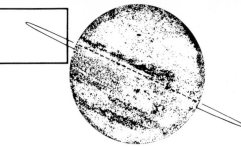

The Jovian system

Four Voyager pictures of Amalthea, Jupiter's innermost satellite and the largest of the swarm of small, rocky satellites which, with the four giant Galileans, make up a family of at least 14 moons. Amalthea is about 280 km long and 150 km across. It keeps its long axis pointed towards Jupiter as it progresses in its 12 hour orbit. The surface material is dark, like a carbonaceous chondritic meteorite, but with a strong red tint which may be due to sulphur contamination. Craters and ridges can be seen, along with bright patches on some slopes due to a green-tinged material of unknown composition.

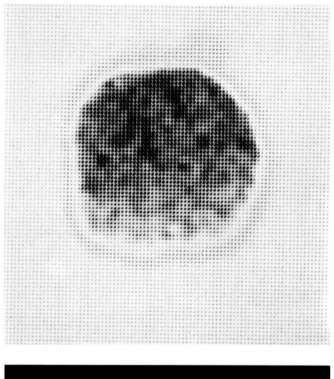

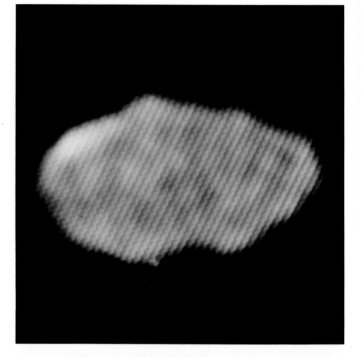

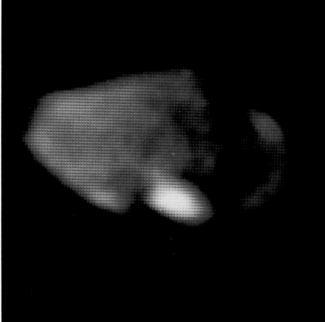

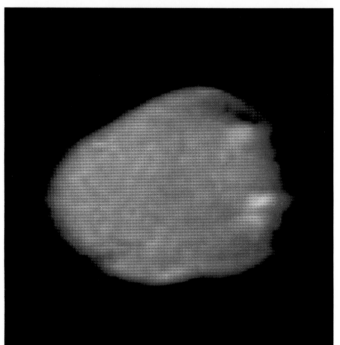

The face of Io. The brown, orange and yellow areas are probably covered by sulphur or a mixture containing sulphur. The white patches may be sulphur dioxide snow, and the pock-marks are mostly volcanic calderas up to 200 km across. Mountainous regions exist near both poles, with some features rising 8 km or more above their surroundings.

Io in transit across Jupiter.

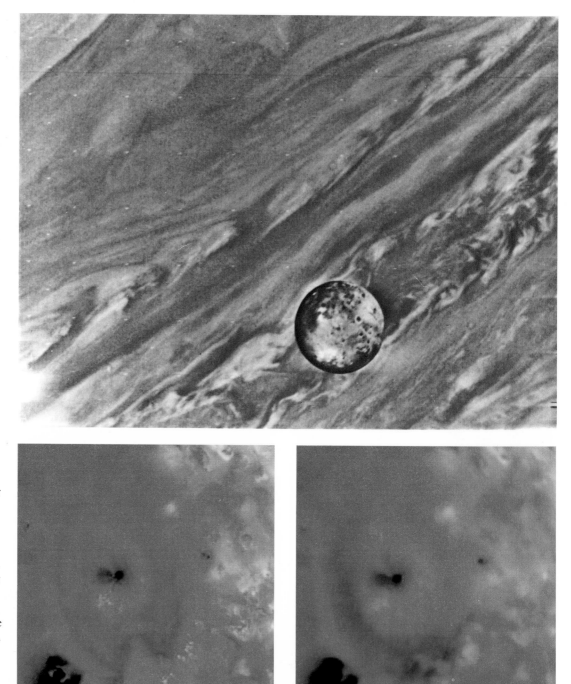

Close-up views of calderas (volcanic craters) on Io. The two pictures were taken four months apart, during which time the appearance of the areas around the eruptions had noticeably changed. This shows how quickly the surface of Io can be covered over by ejecta, and explains the absence of visible impact craters.

A volcanic plume on Io showing the umbrella shape which suggests particles in a ballistic trajectory. With this assumption and the height of the plume (100 km), velocities of ejection can be worked out. These are in excess of 3200 km per hour.

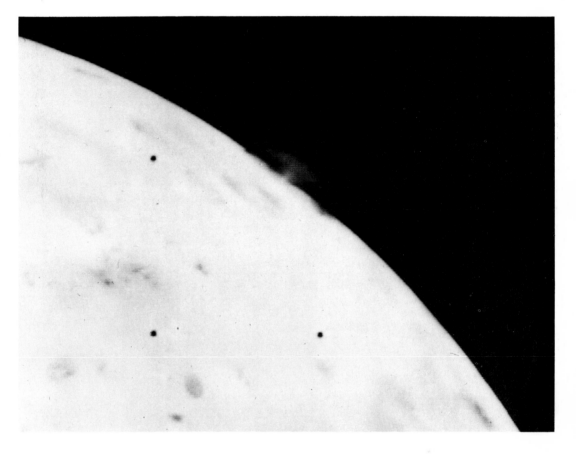

Volcanic activity on Io. These particular volcanoes were seen erupting by both Voyagers 1 and 2, and so presumably had been continuously active for at least four months. One volcano was active for Voyager 1 but quiet for Voyager 2.

Three of the eight active volcanoes so far detected on Io.

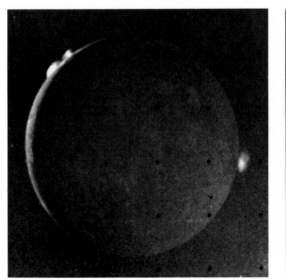

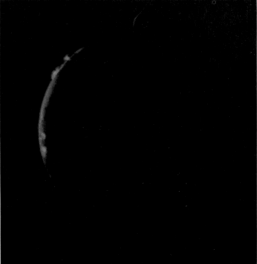

Surface erosion reminiscent of that caused by flowing liquid is seen in this computer-enhanced picture of Io. Subsurface flow is more likely than surface rivers because of the low temperatures (-150 to $-200°C$) and absence of a significant atmosphere. However, conditions may have been different in the past. The diagrams and characters around this image record how the contrast has been manipulated to enhance the details. Also given are details of the exposure etc., and the co-ordinates which allow the features to be located and related to other images. Photographs are most commonly used in this form by scientists in their analyses of the features which they show.

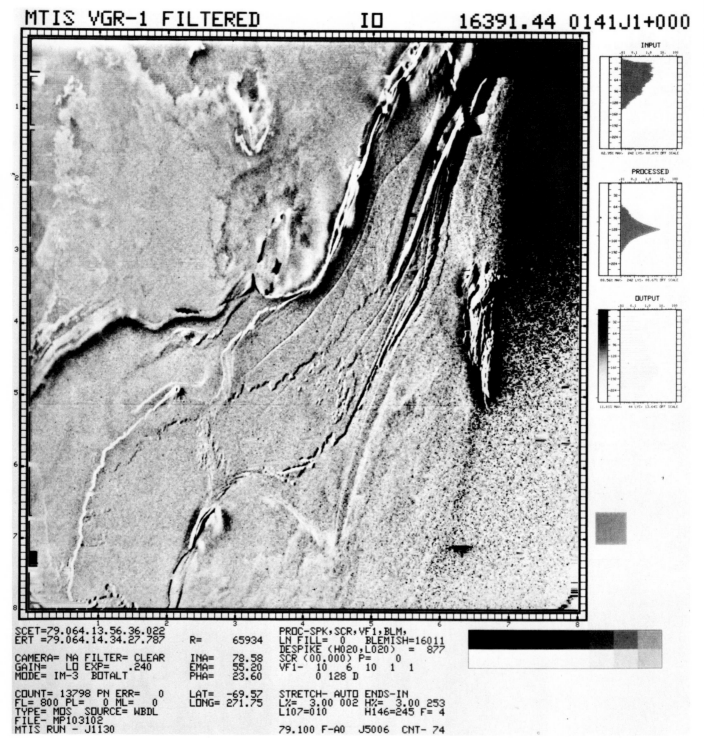

The small bright feature at the upper right is probably an example of a cloud, perhaps of SO_2 crystals, forming over a caldera on Io.

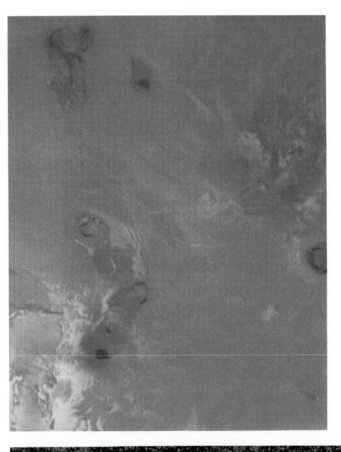

A false-colour image of an eruption on Io. The yellow core is the portion which is seen in visible images; the blue 'halo' is detected only in ultraviolet pictures. The latter is 320 km across, and probably consists of very small particles, perhaps sulphur dust.

Map of Io prepared from Voyager 1 and 2
photographs. 1 degree equals 32 km.

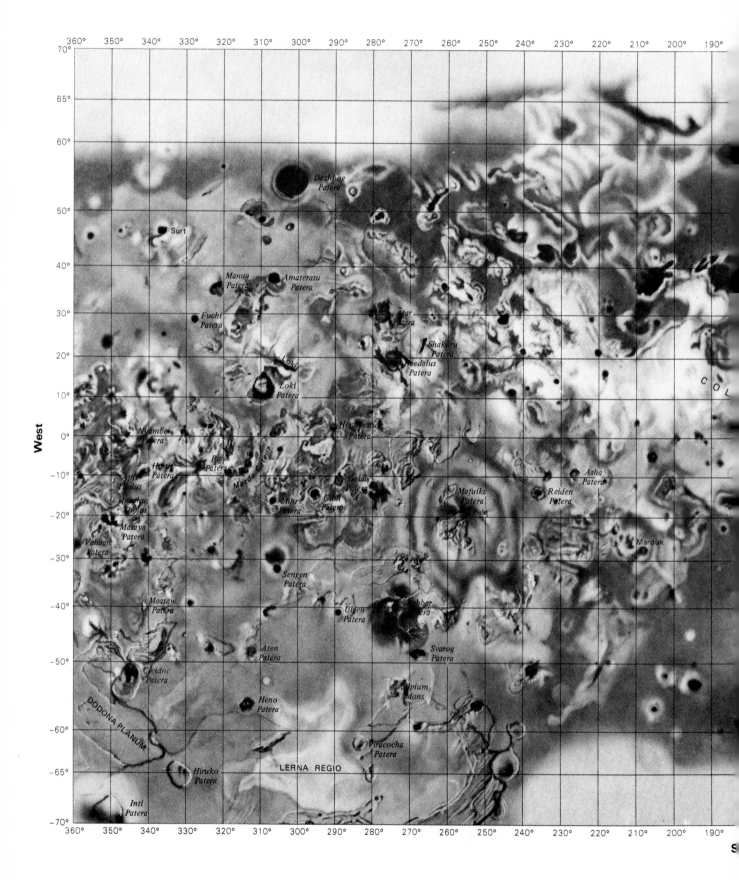

The Jovian system

This close-up view of Io is centred near 72S latitude on the Jupiter facing side at longitude 326°W. The frame is about 1760 km in width and features as small as 5 km can be resolved. Many kinds of typical Ionian landforms are visible on the smooth volcanic plains, including calderas, complex scarps and rugged mountains. The albedo patterns are doubtless associated with the unusual nature of the sulphurous volcanic processes that are unique to this body. Very few of the calderas are associated with edifices having any large vertical relief and as such they differ markedly from those found on the Earth and on Mars. There *is* some relief, however, since flows downhill from the calderas are observed in the pictures.

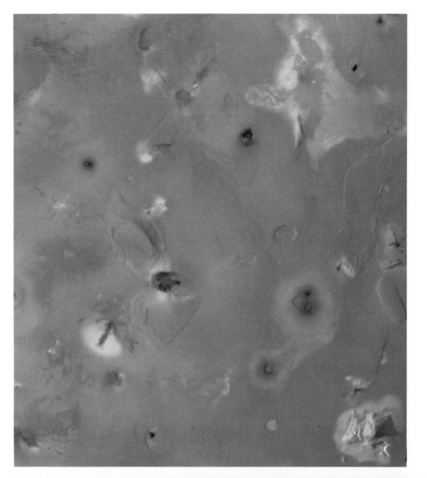

This Voyager 1 view of the limb of Io provides a sense of the rugged terrain encountered in some regions of this moon.

Europa is a very reflective body, owing to its surface covering of ice. If placed where our Moon now is, it would outshine the latter by nearly ten times even though it is about the same size. Also, because of the ice covering, there is very little surface relief; Europa may be the smoothest body in the Solar System.

Markings on Europa consist of two main types: a mottling effect, possibly due to the rocky surface showing through relatively thin ice, and long, linear fissures. The latter have been likened, in distribution over the globe and actual physical character, to the canals of Mars as popularly imagined in the nineteenth century. On Europa, however, the origin is cracking of the satellite's thin, icy crust.

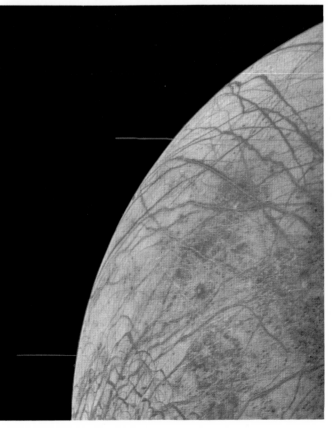

Map of Europa prepared from Voyager 1 and 2
photographs. 1 degree equals 27 km.

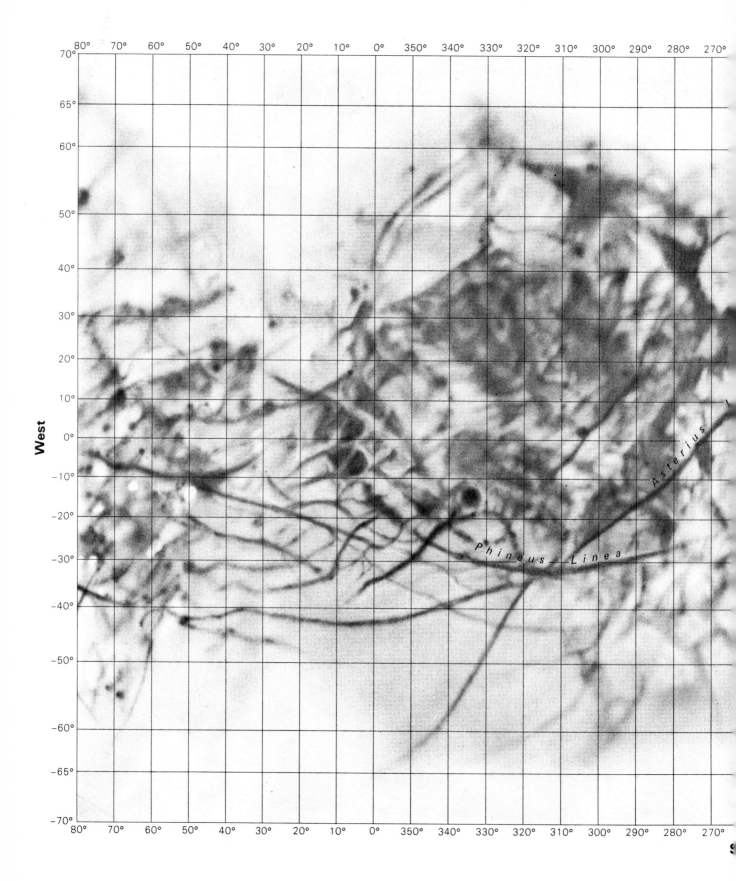

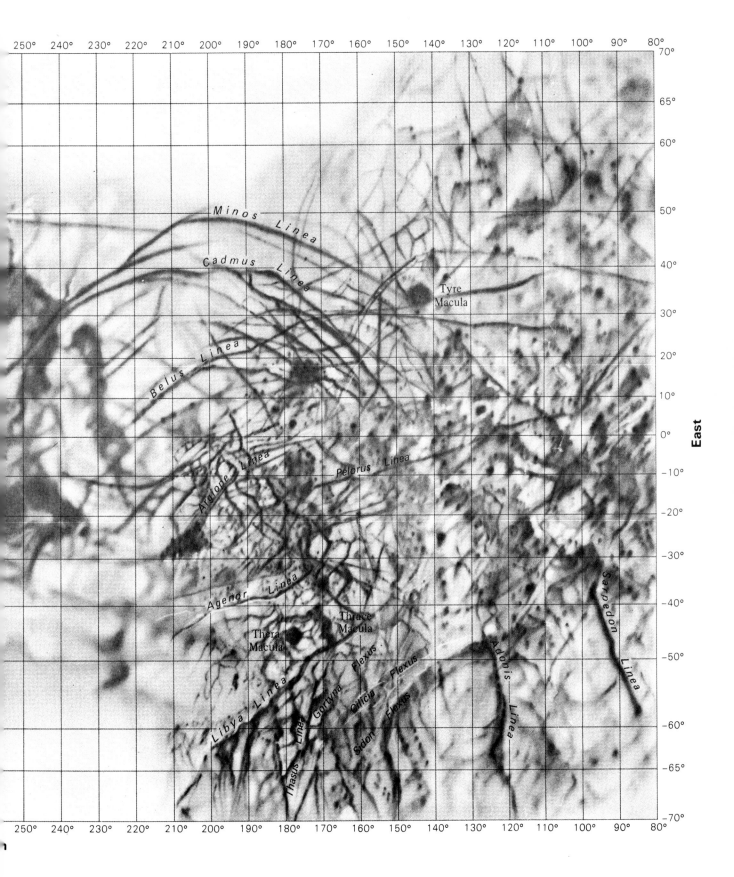

This view of Europa covers a large area of the satellite's surface at close to the highest resolution yet obtained – about 5 km. It shows very well the smoothness of most of the terrain, and the near-absence of sizeable impact craters. Only three craters larger than 5 km diameter have been found anywhere on Europa. Both facts are probably explained by glacier-like 'cold flow' in the icy crusts, which smooths out surface relief and craters over periods of millions of years.

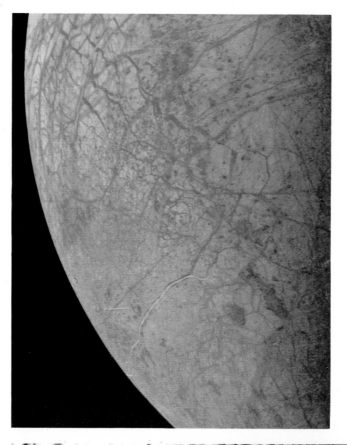

Images of the terminator region are useful for estimating the height of features by measuring the length of shadows which they throw. These ridges on Europa are only about 100 m high. Note the circular feature at the centre of the picture, on the terminator. This is probably the remains of an impact crater, one of very few seen on Europa.

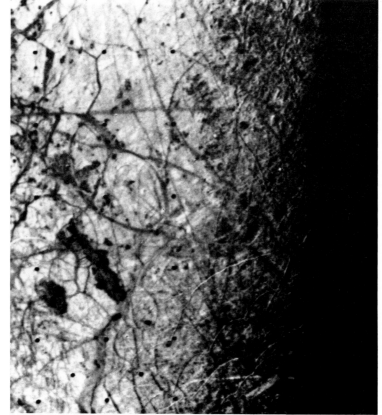

An image of Ganymede showing most of the hemisphere which faces permanently away from Jupiter. The surface, and indeed the interior down to a depth of several hundred kilometres, is mostly water ice. The darkest regions are the oldest, because of the original mixture of rocky and icy constituents and undergoes physical and chemical changes (some due to impurities) during long exposure to space and the Sun's rays. The reason for the survival of a continent of very dark, and therefore ancient, terrain in the Northern Hemisphere, is unknown.

Very large craters, caused by impacts with big bodies like asteroids or comets, are sometimes surrounded by concentric rings for a considerable distance. This example on Ganymede fills an area 1300 km across. The rings were produced by rippling of the crust as it responded to the collision. This event on Ganymede happened several billion years ago, because the central crater has been obliterated by 'cold flow' of the crust over the ages. Other examples of smaller crater 'ghosts' can be seen in this picture, along with more recent craters which still show some relief.

Map of Ganymede prepared from Voyager 1 and 2
photographs. 1 degree equals 46 km.

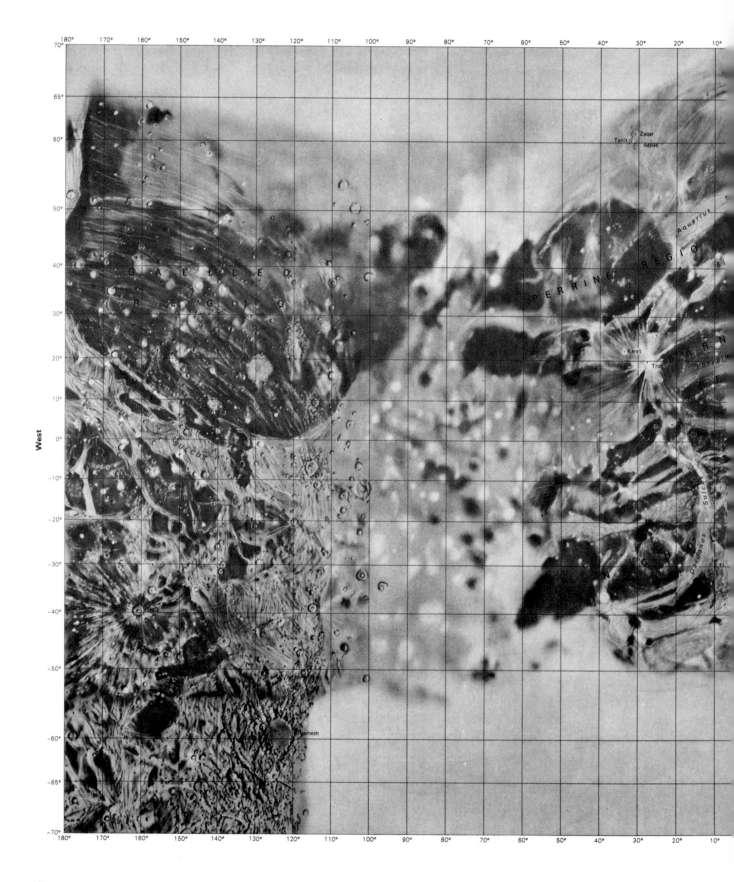

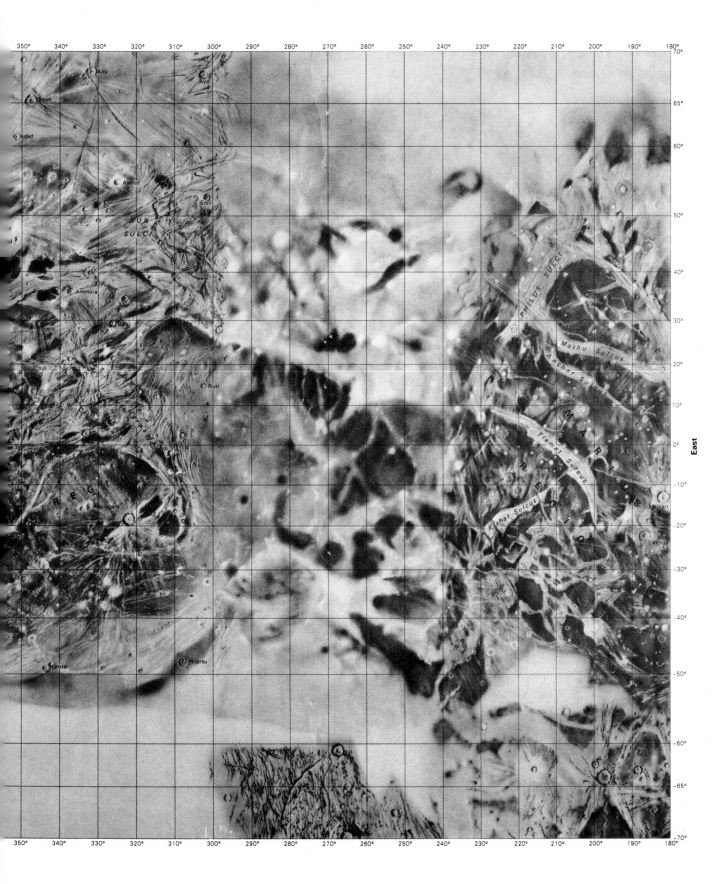

350° 340° 330° 320° 310° 300° 290° 280° 270° 260° 250° 240° 230° 220° 210° 200° 190° 180°

East

Impacts with the icy surface of Ganymede produce
craters, and spray clean ice from below the surface
across the landscape. Fault lines also develop, and
shear along the length of these produces parallel
grooves like the two large examples in the top half of
this picture.

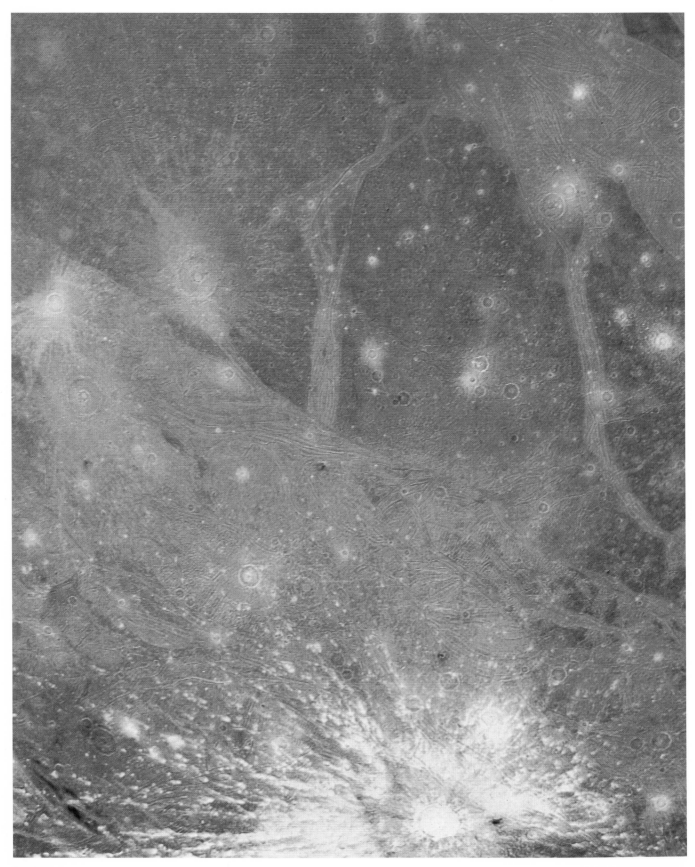

Complex grooved terrain on Ganymede observed near the terminator. The crust must have been much more mobile than it is now, to produce features intertwined in this way. The present temperature of the surface is about −150°C on the sunlit hemisphere, less on the night side.

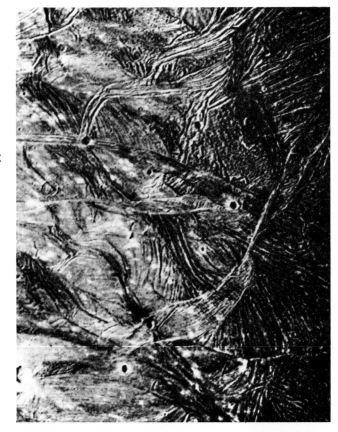

Dark 'continents' on Ganymede, several hundred kilometres across, are separated by lighter-coloured fault regions. Under close examination, features which were once continuous across fault lines can be seen to have been displaced by a process akin to continental drift on Earth.

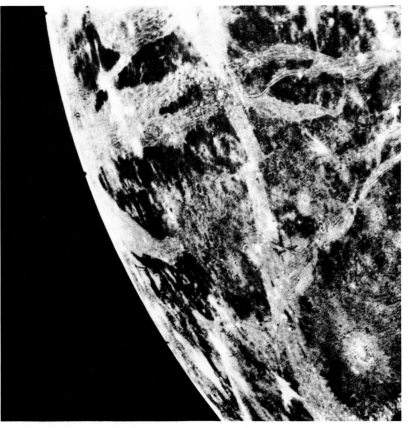

A computer-enhanced mosaic of Voyager pictures, showing the heavily cratered face of Callisto. The density of craters is higher than for any other body in the solar system. Overall, the body is extremely smooth, with no significant mountains or valleys.

Details of part of the concentric ring structure visible in the preceding figure. Note the decrease in crater density towards the impact basin to the left, implying that the event covered over some old craters. However, in some places there are young craters on top of the rings. Apart from impact events, the surface of Callisto appears to have been unmodified for a much longer period of time than the other Galilean satellites, perhaps longer than any other planetary body in the Solar System.

A giant impact basin on Callisto. The central bright
region is 300 km across, and the ring-shaped fracture
lines around it extend for 1500 km. The impacting
body – a large asteroid, or a comet – may have
gone right through Callisto's icy crust into the
subterranean ocean below.

Map of Callisto prepared from Voyager 1 and 2
photographs. 1 degree equals 42 km.

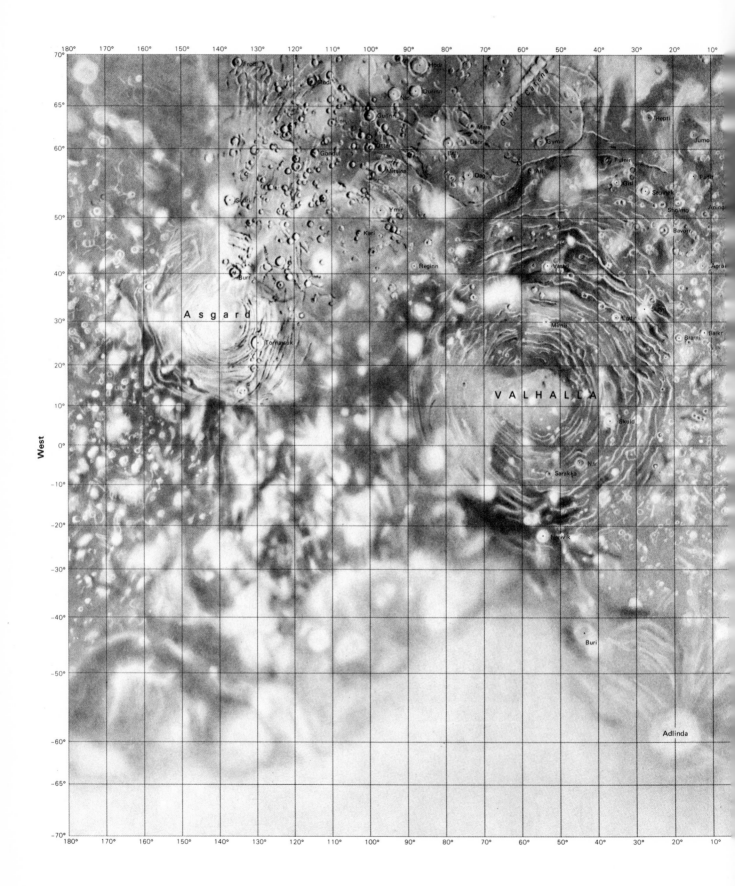

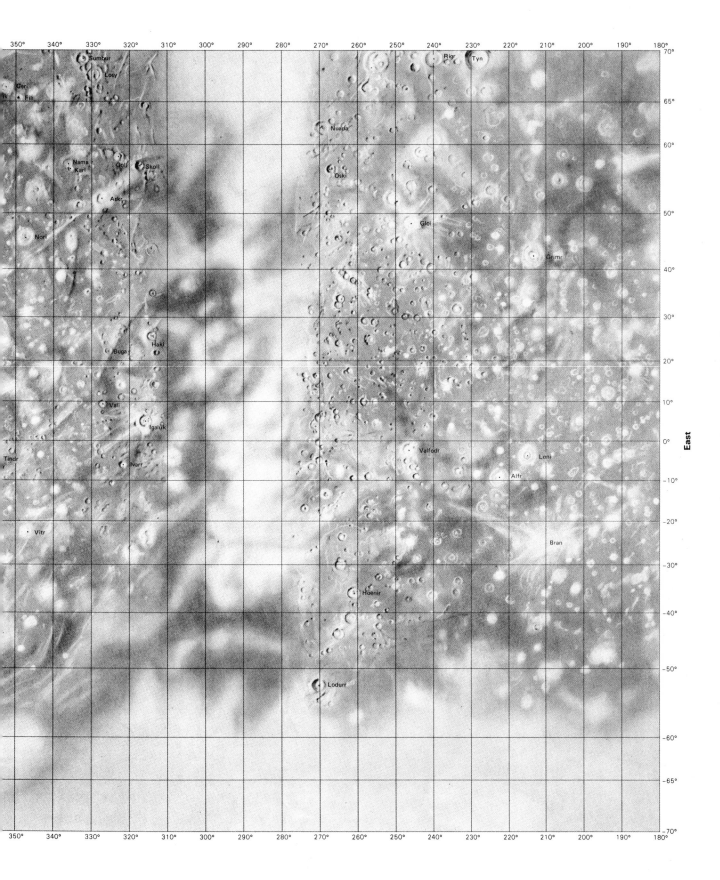

The Saturnian system

Saturn The ringed planet Saturn is arguably the most beautiful object in the known universe. The combination of aesthetic appeal with the numerous scientific mysteries associated with this distant world placed the Voyager explorations in 1980–81 among the most eagerly anticipated since space flight began. In addition to investigating the nature and origin of the ring system – thought until recently, to be unique in the Solar System – the planet's atmosphere and family of moons were targets for Voyager's cameras. The atmosphere is in the same class as Jupiter's, very deep and consisting mostly of hydrogen, but different in many important details, such as minor constituent abundances, internal heat source and distance from the Sun. The satellites are numerous and mostly icy, like Jupiter's, but they formed in a different part of the primordial nebula (and so probably do not have quite the same mix of elements) and exist today in a significantly colder environment. They should, therefore, be different in character in ways that reveal themselves in high-resolution pictures. Let us look in more detail at the atmosphere, the rings and the satellites of Saturn, in turn. Some statistics about Saturn are summarized in the following table:

Mean distance from Sun 9.54 AU	Polar radius 49 000 km
Orbital period 29.5 years	Mass 5.69×10^{26} kg
Inclination of equator 29°	Mean density 0.69 gm/cm³
Equatorial radius 60 000 km	Period of rotation 10 h 40 m 30 s

The resolution in pictures obtained as Voyager approached Saturn exceeded the best ever obtained from Earth, even when the craft was still over 30 million kilometres away. The globe is visibly flattened at the poles, a result of the very rapid rotation of the planet on its axis. The banded appearance is subdued compared to Jupiter, but unmistakably present; in 'stretched' images, where the contrasts between different colours have been exaggerated, the bands show up very clearly. The nomenclature for the bands is based on that used for Jupiter.

The reason for the subdued contrasts in the cloud bands on Saturn, relative to Jupiter, is not certain. Saturn has a higher ratio of internal to external (solar) atmospheric heating than Jupiter – about 2:1 compared to about 1:1, as revealed by measurements of albedo (reflected) versus thermal (emitted) fluxes. This might be expected to produce more vigorous convective activity and hence more defined cloud bands than on Jupiter. One possibility – that the contrasts lie below an obscuring stratospheric haze – seems to be ruled out by the fact that such features as can be seen appear just as sharply near the limb as in the centre of the disc. The limb features are viewed through a longer path of overlying atmosphere and would be more heavily obscured by any haze above the tropopause. A likely, but unproven, explanation is simply that the belts on Saturn are cloudier than those on Jupiter because they are colder. The tropopause temperature on Saturn is nearly 200°C below zero, so ammonia starts to condense much further (actually about 100 km further) below the tops of the convection cells than it does on Jupiter. On the warmer planet, ammonia starts to condense just a few kilometres below the point where the transition to downward motion starts, so the ammonia clouds are only a few kilometres thick in the (rising) zones and can be completely cleared in the (descending) belts. On Saturn it appears that the much thicker ammonia clouds are not completely cleared in the belts, even though the downward motion may be more rapid.

In addition to the reduced belt–zone contrast, discrete features such as large ovals of any colour appear to be much less visible on Saturn than on Jupiter. By extreme contrast enhancement of the Voyager pictures, it has nevertheless proved possible to observe a few long-lived spots and plumes. The scale and appearance of these, and their movement and evolution in successive images, provide clues to the circulation and meteorology on the planet. The details seen include Jupiter-like red and brown spots, convective clouds on scales considerably smaller than the belt-zone separation, and wave patterns of various kinds. Feature tracking has revealed unexpectedly high velocities in the equatorial band, about five times more rapid, at 1800 kilometres per hour, than Jupiter's rapid equatorial jet. At present, it is not clear which aspects of the differences in structure between the two atmospheres force this large difference. Perhaps the most likely possibility is that the jets on both planets are accelerated by the energy carried by eddies convected up from below. The greater tendency towards convection on Saturn caused by the larger internal to external heating ratio may drive the higher velocity. Of course, conditions of temperature, pressure, gravity, and perhaps composition are different on the two planets also. In particular, we observe the zonal winds at only one height level on each planet and in this situation it is impossible to be completely diagnostic. For example, the winds on Jupiter may be considerably higher than those on Saturn at some level above or below that observed. So we must be careful not to stretch any comparison too far until we have vertically resolved measurements from some future probe or orbiter. Nevertheless, to explain the remarkably high winds on Saturn at any level near about half an atmosphere of pressure is a stringent test of any theory of atmospheric dynamics.

Before Voyager, Saturn's rings were considered one of the great mysteries of the Solar System, and since obtaining detailed data on them the problems have, without doubt, increased. Firstly, their origin is obscure. They may have condensed in their present form or they may be made up of the debris resulting from the break-up of a satellite. Their composition is not known with any certainty, but their infrared spectra do show the signature of water ice. The rings could be composed principally of icebergs and/or snowballs ranging in size from a centimetre or so to several metres in radius. Their very low emissivities at radio wavelengths and high radar reflectivities argue against a significant content of silicates.

The observed radial distribution of the three major rings (see the following table) is due, primarily at least, to gravitational interactions with the satellites:

Major satellites and rings of Saturn

Name	Distance from Saturn (planetary radii)†	Width or diameter (km)
C ring	1.37	19 000
B ring	1.74	25 200
A ring	2.16	15 600
Mimas	3.09	390
Enceladus	3.97	500
Tethys	4.91	1050
Dione	6.29	1120
Rhea	8.78	1530
Titan	20.4	5118
Hyperion	24.7	310
Iapetus	59.3	1440
Phoebe	216.0	150

† Equatorial radius of Saturn $= 6 \times 10^4$ km.

The earlier attempts to quantify this effect focused on Mimas, until recently the closest of the known satellites to the rings. It was anticipated that orbits with periods equal to proper fractions of the orbital period of Mimas would be unstable, i.e. would contain no particles. The reason for this is that the gravitational attraction of the larger body tends to tug the smaller one out of its orbit. This is unimportant if the perturbation occurs in uncorrelated directions as a function of time, but for orbits in resonance the perturbation tends to accumulate. This produces an increasingly more elliptical orbit for the ring particle, until eventually it collides with another and resumes in some other orbit altogether. In support of this mechanism, observations show that the inner and outer edges of the most prominent ring (the B ring) lie at distances corresponding respectively to periods equal to $\frac{1}{2}$ and $\frac{1}{3}$ of that of Mimas. The encouragement given by this example is partially offset by the fact that the detailed theory of orbital resonance predicts a width for Cassini's division of less than 30 km, more than 100 times too small. However, this difficulty became less relevant when the true complexity of the ring structure became apparent from the first Voyager images, in which approximately 95 concentric rings were visible. Later images, with still higher resolution, put the count into thousands, while photopolarimeter observations of a star occulted by the rings raised it to hundreds of thousands. Furthermore, the colours of the ringlets and their variable opacity to radio waves suggests that they are sorted by particle size and possibly also by composition. Even though Saturn is now known to have at least sixteen satellites in a variety of orbits (see below), it remains to be shown that the resonance theory alone can account for the complexity of the ring structure as it is now revealed.

The principal new features of the ring system, discovered by Voyager, are as follows. In addition to the multiplicity of 'ringlets' within the major systems, material forming discrete ringlets is found, within Cassini's and Encke's divisions. When viewed from below (i.e. with the rings between the spacecraft and the Sun), Cassini's division appears brighter than the adjacent A and B rings. The C ring behaves similarly. Known also as the 'crepe ring' for its tenuous appearance through Earth-based telescopes, it appears brighter than the relatively thick B ring from behind. The reason is partly because of the scattering properties of the particles in the rings, and also because the B ring is thick enough to be nearly opaque. This is evident because the disc of Saturn is partially obscured by the B ring when viewed through it.

In addition to the concentric divisions, Saturn's rings also contain radial features, or 'spokes', discovered by Voyager. These can be seen clearly in a series of photographs taken at different times. These features rotate with the rings and tend to be tilted with respect to the radial direction, due to the fact that the ring particles themselves must rotate with gradually increasing periods with increasing distance from Saturn along a spoke. It has been suggested that the smallest ring particles are acquiring an electrostatic charge by interaction with the planet's magnetic field, and repelling each other with electromagnetic forces which overcome the gravitational attraction. To this simple picture must be added some, as yet unknown, mechanism for the localized and time-dependent nature of the phenomenon. Perhaps the rings are responding to 'magnetic storms' within Saturn which produce transitory enhancements of the field.

Outside the main ring system, there exists a very narrow ring less than 150 kilometres in width. This was discovered by Pioneer II, and named the F ring. It is accompanied on either side by two small satellites, each about 250 kilometres in diameter. Their presence will tend to focus the particles in the F ring, the outer satellite, by trailing, hence decelerating, the outermost ring particles and moving them inwards, and the converse for the inner satellite. This effect probably explains the ring's narrowness. What it does not explain, however, is its detailed structure. The ring contains at least five components, moving in irregular orbits which cross each other in places. The analysis of such a behaviour requires theories of orbital dynamics of significantly greater complexity than exist at present.

The moons of Saturn

We turn now to Saturn's family of satellites. Some properties of these are summarized in the following table.

	Radius (km)	Mass (gm)	Density (gm/cm^{-3})	Distance from Saturn (km)	Orbital Period (days)
Mimas	195	3.76×10^{22}	1.2	185 000	0.942
Enceladus	250	7.40×10^{22}	1.1	238 000	1.370
Tethys	525	6.26×10^{23}	1.0	295 000	1.888
Dione	560	1.05×10^{24}	1.4	377 000	2.737
Rhea	765	2.28×10^{24}	1.3	527 000	4.518
Titan	2560	1.36×10^{26}	1.9	1 222 000	15.945
Hyperion	160×100	1.11×10^{23}	—	1 481 000	21.277
Iapetus	720	1.93×10^{24}	1.1	3 560 000	79.331
Pheobe	70	—	—	12 930 000	550.40

Mimas

The innermost of the nine satellites known since early times is Mimas, which was discovered by Herschel in 1789. Its role in producing the major divisions in the rings was noted above. Voyager 1 obtained the first pictures to show any detail of the surface. These reveal a bright, icy terrain, heavily cratered. In view of its low density, the satellite probably consists mostly of ice. The lower insolation at Saturn's distance from the Sun, compared with Jupiter's, results in substantially lower surface temperatures and increased rigidity of the icy crust. Probably, flow in water ice at temperatures as low as those on the surface of Mimas (about $-200°C$) is small enough to retain impact features dating back to the time of the satellite's formation. One of the craters on Mimas is spectacularly large, in relation to the size of the body. Discovery of such a high ratio of crater radius to satellite radius (about 1 to 4) leads to speculation about the limiting size of impacts which can occur without disintegrating the satellite altogether. Traces of fracture marks on the opposite side of Mimas testify to the fact that this event must have come quite close to this limit.

Enceladus

Enceladus is a near-twin to Mimas in terms of size but appears to have a smoother, brighter surface. It had been expected that both were too small to show evidence of surface activity caused by internal heating, but Voyager photography of Enceladus has revealed fissures, plains, corrugated terrain and other evidence of crustal deformation. It is now postulated that the interior of this moon is heated by a tidal mechanism similar to that which melts the interior of Io. Enceladus is deformed rhythmically by Saturn's gravitational field as its orbit is perturbed by the much larger moons Tethys and Dione nearby.

Tethys, Dione and Rhea

Tethys, Dione and Rhea are also icy bodies of various densities, the variations probably indicating different mixes of ice and rock. The value of 1.4 ± 0.1 g/cm^3 for Dione, for example, is consistent with a rocky core making up about one-third of the satellite's mass. Tethys, on the other hand, has a density of 1.0 g/cm^3 and must be nearly all water. All are heavily cratered, and show surface fissures and interesting albedo contrasts. In most cases, the latter can be attributed to the age of the terrain, with brighter regions covered by fresher ice thrown up as a result of crater formation by impacting debris. Others look more like sinuous deposits of snowy material, possibly formed by liquid or vapour from below seeping up through cracks in the crust.

All three satellites are brighter, on average, on the hemisphere which leads the satellite around its orbit than the other. Also, the trailing hemispheres generally show light and dark markings which appear to have been mostly obliterated on the leading side. The presumption is that the latter has been subjected to 'gardening' by debris from outside the Saturnian system while the trailing edge remained relatively protected.

Titan

Titan is the grandest of Saturn's satellites. With its great size, just slightly smaller than Ganymede, and its thick atmosphere, it physically resembles one of the terrestrial planets more closely than it does any of the small, icy satellites of Saturn. However, its low density rules out the possibility that it is a totally rocky planet, like Mars for example. Probably it has a silicate core, about half the volume of the visible globe, overlaid by a thick aqueous crust and topped by a deep atmosphere. Calculations indicate that, in the absence of much tidal or radioactive heating, Titan is likely to have lost most of the original heat of accretion which differentiated its interior and become cool throughout. The aqueous mantle, therefore, probably froze long ago and now consists of solid ice. Models of the atmosphere, strengthened by Voyager radio occultation measurements, show a depth of about 200 km from the surface to the visible limb, and a surface pressure of around 1.5 Earth atmospheres. Ultraviolet and infrared spectroscopy reveals the composition to be 85% nitrogen, probably produced by photodissociation of ammonia with subsequent escape of the resulting hydrogen. The remaining 15% is mostly argon, with about 1% of methane, which is the gas whose spectral signature was first observed by Kuiper in 1943, leading to the original discovery of an atmosphere on Titan. Partly because it was not known that methane is a minor rather than a major constituent, the great depth and substantial surface pressure of Titan's atmosphere was not appreciated until improved spectroscopy was performed in the early 1970s. Then it also began to be suspected that Titan's surface is completely obscured by cloud, rather like Venus. In the case of Titan, the cloud is probably condensed methane overlaid by a haze of photochemical 'smog'. The latter can be seen in close-up photographs of the limb. It is presumed to be produced from methane by solar ultraviolet radiation and charged particles from Saturn's radiation belts, which dissociate the CH_4 and allow the formation of higher hydrocarbons, which form oily droplets. Alternatively, the haze may be wholly or partially dust, or some other material altogether. In any case, photochemical activity clearly is important in Titan's upper atmosphere because ethane, ethylene, acetylene and hydrogen cyanide – the last mentioned in an abundance of only about 3 parts per million – are all observed spectroscopically. Polymerized cyanides have an orange colour which may account for the hue of Titan's clouds. Polyacetylene is another candidate which has been suggested. The production of aerosol probably depends sensitively on the solar intensity, accounting for the difference in brightness between the northern (winter) hemisphere and the brighter southern hemisphere at the time of the Voyager encounter.

The total thickness of the cloud layers on Titan is about 200 km. It is hypothesised that the droplets will grow until large enough to drizzle down on to the surface. Such a process will gradually deplete the methane content of the atmosphere unless a reservoir exists somewhere.

In fact there may be several such reservoirs – at about $-180°C$ the surface of Titan is cold enough to have lakes of liquid methane, while methane clouds in the troposphere are also possible. Solid methane and ammonia ices may be mixed with the surface water ice and the ammonia would then provide a source of fresh nitrogen to replace any lost from the top of the atmosphere by collision with energetic ions from the Saturnian radiation belts. The vapour pressure of water is so low at $-180°C$ that virtually no water vapour will escape into the atmosphere, even though the latter is probably in intimate contact with vast amounts of solid water. It is now generally accepted that early speculation about Titan as an abode of extraterrestrial organisms was premature since liquid, or at least gaseous, water is an essential ingredient for life as we can conceive it. The crucial unknown about the surface of Titan at the present time is whether it consists of the icy crust exposed to the atmosphere, a planet-wide methane sea or (as some calculations predict) a layer of precipitated hydrocarbons several hundreds of metres thick.

Hyperion

Hyperion is a small, pock-marked body of irregular shape. Its most curious feature is its orientation, with its major axis tilted at 45° to the plane of its orbit in the available Voyager 2 pictures. Such a position is unstable and a possible implication is that Hyperion is tumbling along its orbit. This would imply that the satellite was captured relatively recently, or that it had undergone a relatively recent major collision.

Iapetus

The most distinctive feature of Iapetus is the very great contrast in reflectivity between its two hemispheres. Unlike Tethys, Dione and Rhea the *leading* side on Iapetus is dark, and slightly reddish in colour. With an albedo of only 4% (compared to 50% for the trailing hemisphere) the material responsible is one of the least reflecting substances known, perhaps soot or a dark porous mineral like that constituting carbonaceous chondritic meteorites. One theory is that the crust of Iapetus contains an abnormally large fraction of methane ice, some of which has been converted by some means into carbon. Alternatively, Iapetus may have been subjected to an unusually large flux of dark meteoritic material from space. One feels the need to be cautious with the latter explanation, since it is similar to that invoked to explain contrasts on Tethys, Dione and Rhea which are in the opposite sense!

The planet Saturn, as photographed on 4 August 1981 by Voyager 2, from a distance of 21 million kilometres. Many features can be seen, for example, the pronounced flattening of the globe caused by the rapid spin rate (about one rotation every 10 hours) and the fact that Saturn is gaseous to a great depth and not a rigid body. The faint streaks across the disc are cloud bands similar to Jupiter's, but less well defined because the clouds condense at deeper levels in Saturn's colder atmosphere. The rings were named before it was realized that there are more than three of them. The classical A ring is the outermost, actually seen here as a double system with Encke's division appearing near the outer edge. Cassini's division separates the A ring from the B ring, which is the widest and densest of the three. Inside the B ring is the very tenuous C ring, which was discovered by Earth-based observers in the mid nineteenth century. It is also commonly known as the 'crepe' ring. Three of Saturn's moons (Tethys, Dione and Rhea) also appear and the dark spot on the southern hemisphere is Tethys's shadow. Mimas is in transit across the disc and can just be seen south of the rings.

In these views the colour contrasts have been computer-enhanced to bring out details in the cloud bands. The predominance of blue in the southern hemisphere is due to increased molecular (Rayleigh) scattering caused by the more oblique viewing direction. Colour contrasts in the northern hemisphere are due to differences in the structure and/or composition of the cloud bands. The main constituent of the visible clouds is probably ammonia crystals, with unknown impurities. The latter probably include various sulphur and phosphorus compounds, which may account for the yellow-brown appearance of the planet to the eye (cf. preceding figure). The two views were obtained nine months apart by Voyagers 1 and 2. The biggest change is in the brightness of the rings – they were tilted more towards the Sun for Voyager 2 (bottom). Subtle changes in the clouds on the planet can also be seen.

Some details of the cloud bands on Saturn, showing similarities with Jupiter, including two brown ovals. The different shades of the bands are probably caused by differences in thickness and particle size of the cloud layers, and by different amounts and types of impurities such as sulphur and phosphorus compounds (or even the elements themselves, in very small amounts).

Cloud bands in the northern hemisphere of Saturn, with the colour contrasts enhanced to bring out details. Note the ribbonlike wave embedded in the light-coloured zone, which is travelling east at 150 m/s relative to the dark (actually brown) oval feature below. The convective clouds below the oval are travelling west at 20 m/s. Rotating eddies like this one transfer energy to the circum-equatorial winds (jets) rather like rollers driving a belt. Two smaller eddies, white in true colour, appear upstream of the large oval.

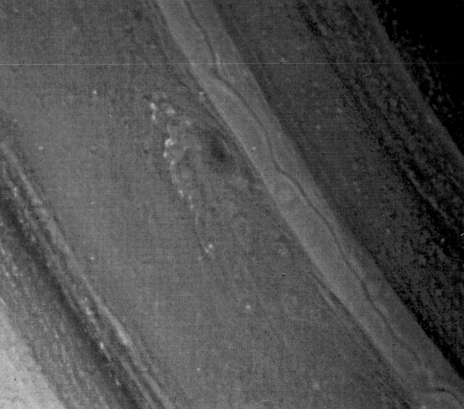

Another example of interaction between a large eddy and a strong eastward jet. The wavelike structure marks an instability in the core of the jet.

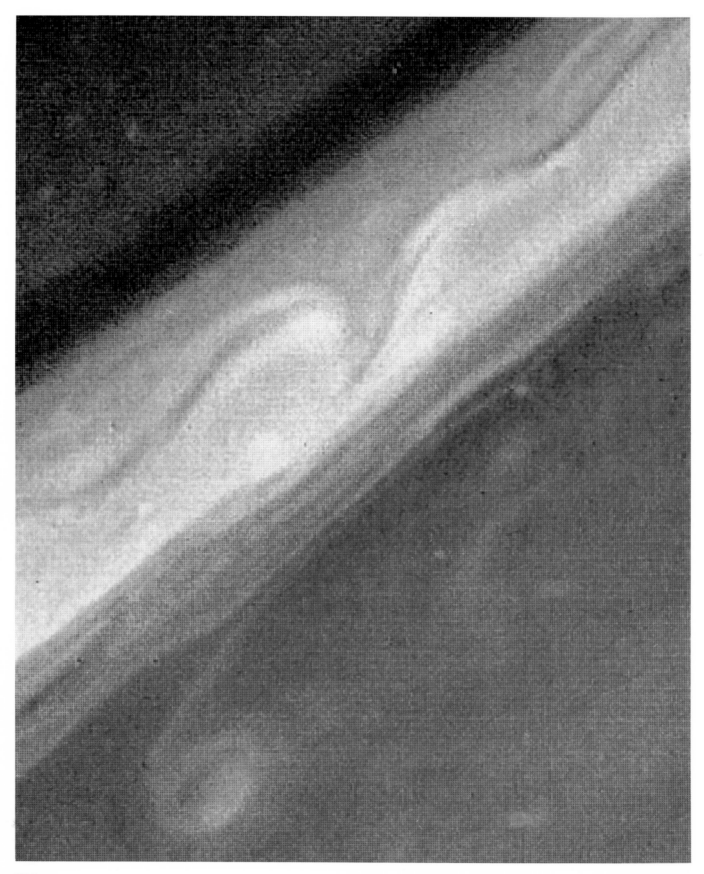

The bright oval feature near the centre of these images is the same giant eddy (about 2500 km across) observed nearly a year apart by Voyager 1 (top) and 2 (bottom). White ovals on Jupiter can last for 50 years or more; Saturn may be similar. In contrast, the less organized, convective disturbances to the north change completely in a matter of days.

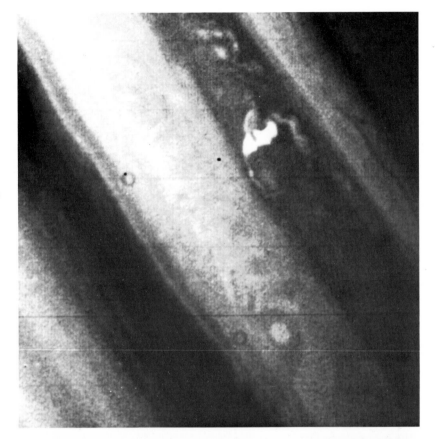

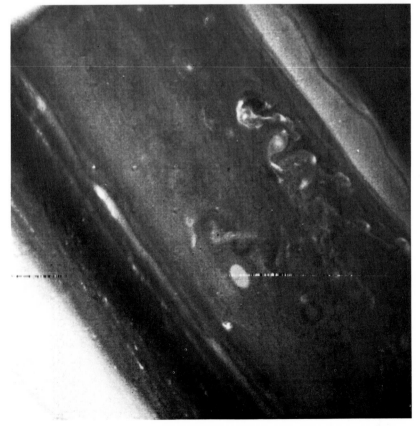

The south polar region on Saturn showing how the relatively bland banded structure breaks up into smaller-scale structure. Apparently the bands are unstable at high latitudes and the atmosphere becomes more turbulent. A particularly well defined wave can be seen at the top left.

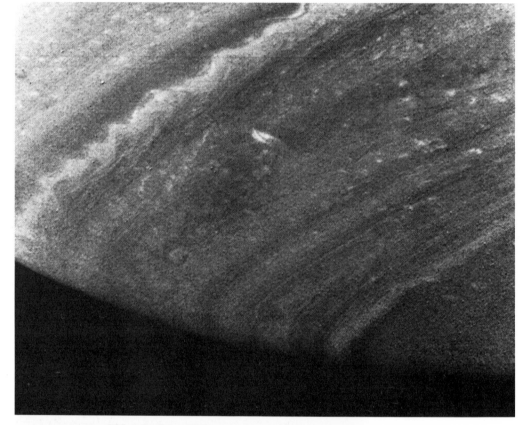

A Voyager 2 image of the north polar region on Saturn showing cloud sheets and three anti-cyclonic (clockwise) eddies each about 250 km across. The two near the middle of the picture are at about 72°N, and the one at the bottom edge at 55°N.

Saturn's three main ring systems can be seen casting shadows on the disc in this view from 13 million kilometres distance. Note that the disc is clearly visible through the crepe ring and through Cassini's and Encke's divisions. The two bright objects above Saturn are Tethys and Dione.

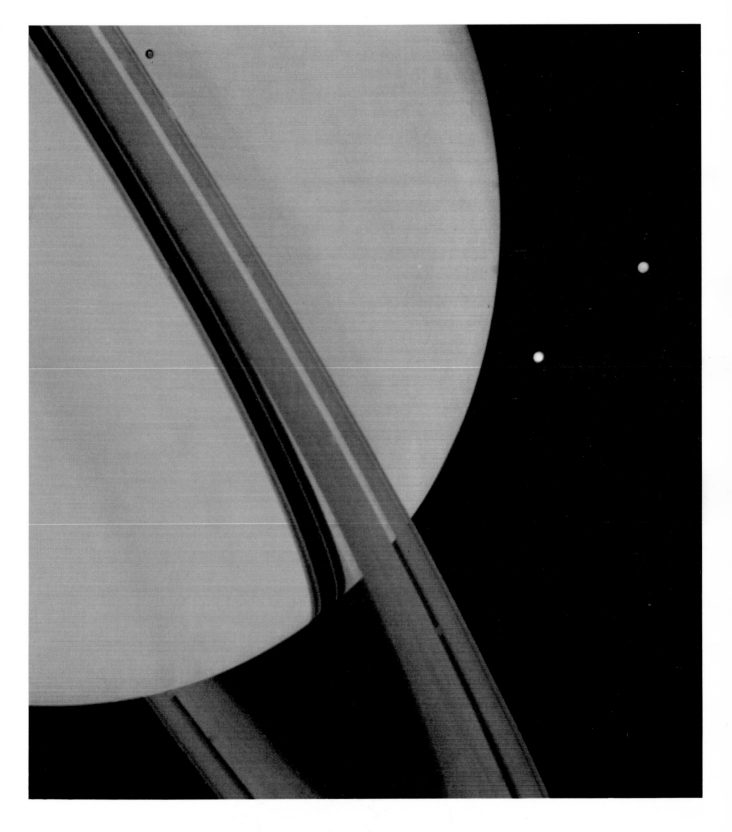

Voyager pictures such as these reveal the complexity of Saturn's ring system and demolish the old concept of three discrete rings. At least 95 individual rings can be counted in this picture.

The narrow, well-separated one at the outside edge is the F-ring; just inside it appears the unnamed satellite S14 which is about 250 km in diameter.

0

The Voyager 2 close-up of part of the B ring shows that even the rings in the previous plate are themselves made up of ringlets less than 15 km across. At the highest resolution obtained so far (about 1 km, by the Voyager 2 photopolarimeter in the stellar occultation mode) these ringlets are found to subdivide still further. Thus, Saturn has at least 100 000 discrete rings and possibly many more.

This exposure brings out the detail in Cassini's Division. Once thought to be an empty space about 5000 km across, it can now be seen to contain material which forms several discrete ringlets. The dark line near the outer edge of the rings (top left) is Encke's Division.

A view of the rings from the dark side, never seen from Earth. Under these conditions the thickest part of the rings (the classical C ring) on the right appears dark, because little sunlight penetrates. The A ring is thinner, therefore brighter, and Encke's Division is brightest of all. Evidently the particles in the Division are relatively small in number and good scatterers of sunlight. They may resemble snowflakes or hailstones made, like their familiar terrestrial analogues, of water ice. The filamentary F ring is also bright when viewed from this angle.

The tenuous C (crepe) ring also shows its detail best in transmitted light. This photograph of the 'dark' side of the rings reveals dozens of ringlets and divisions in this, the faintest of the three classical rings. As in the preceding figure the thicker B ring appears very dark and the intermediate thickness A ring moderately bright. Cassini's Division is nearly as bright as the C ring and the outermost F ring is just visible. Voyager was about a million kilometres from Saturn when this view was obtained.

Although it contains more material than the other regions of the ring system, the B ring region is still transparent as this view of Saturn through the rings shows. The crepe ring region hardly obscures the disc at all. Note also the shadows of the rings on the planet, and the faint illumination of the night side by reflection from the rings.

A colour-enhanced image of the rings exaggerates real colour differences which are thought to be due to differences in chemical composition, although differences in microstructure could also be responsible.

This Voyager 2 picture shows the dark 'spoke-like' features in the rings, thought to be formed by interactions between the smallest ring particles (perhaps dust on the larger particles) and inhomogeneities in the planet's magnetic field. The features form extremely rapidly and have very sharp edges initially; differential rotation rates with distance from Saturn eventually dissipate them.

A close-up view of the F ring. The bright filaments which make it up are not more than about 15 km across. Two small satellites, S13 and S14, orbit just inside and just outside the ring and keep it narrow by 'gravitational focusing'. The irregular shape and intertwining of the orbits of the ring components can be seen quite clearly.

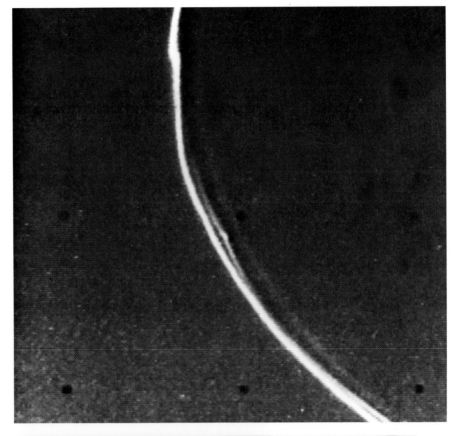

This Voyager 2 image of the F-ring captured the two 'shepherding' satellites when less than 1800 km apart. The inner satellite laps the outer every 25 days; it did so about 2 hours after this picture was taken.

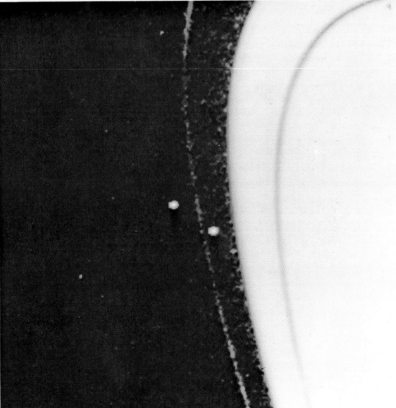

Map of Mimas prepared from Voyager 1 and 2 photographs. 1 degree equals 3.4 km.

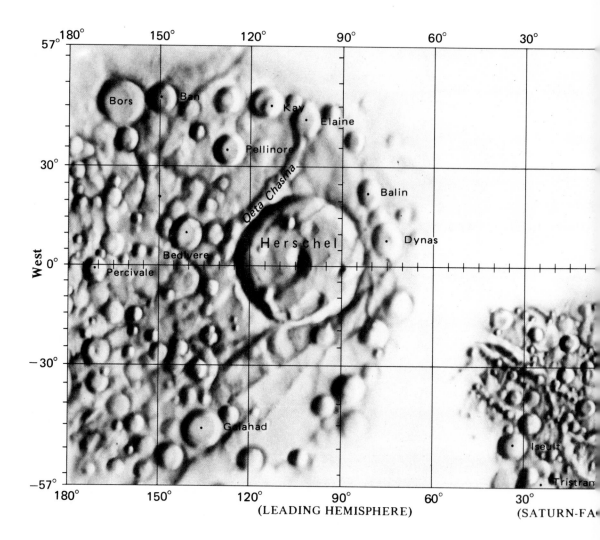

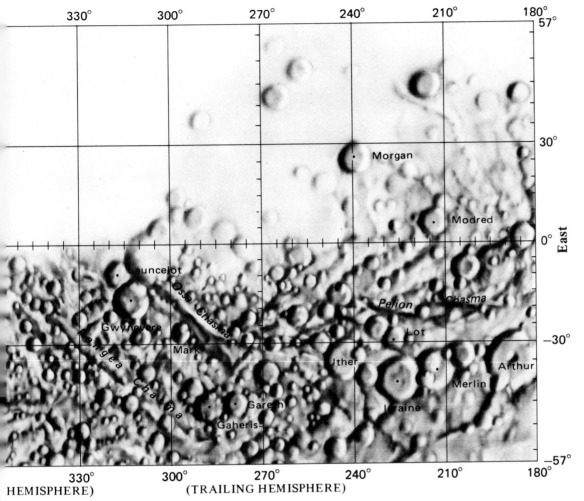

Mimas is the innermost of Saturn's larger moons and measures about 390 km across. The large crater seen here has walls and a central peak nearly 10 km high. It represents the record of a collision which probably occurred some 4 billion years ago. The large size of the crater corresponds to an impact large enough to nearly shatter the planetoid. Saturn's rings may be the remnant of such a shattered moon which used to orbit closer to Saturn.

Mimas is heavily cratered all over its surface, which consists mostly of ice. The smallest craters seen here are about 2 km across. The linear features are probably cracks caused by the large impact on the other side (see preceding figure); their appearance supports the conclusion that Mimas must nearly have disintegrated under the impact.

Voyager 2 got much closer to Tethys than its sister craft did, and obtained this portrait with a resolution of 5 km. Note the difference between the heavily cratered terrain to the top and the much smoother terrain at bottom right; this is evidence of internal activity which resurfaced the younger terrain at some time in the past.

Tethys is Saturn's fifth largest satellite with a diameter of 1050 km. The most remarkable feature of its appearance in this view is the large-scale albedo contrast, with the left-hand side much darker than the right. The impression of two-thirds full phase is caused by this effect; in fact the satellite is nearly fully illuminated in this view, with the terminator on the right. The reason for the distribution, which appears in various forms on several of the satellites, is a complete mystery at present. The feature near the centre of the disc is also interesting. It is about 150 km across but shows little or no relief. Perhaps it is the scar of an impact which occurred early in the history of Saturn when Tethys still had a liquid core, which filled the crater. Today, Tethys is probably solid ice throughout most of its radius with a small rocky core.

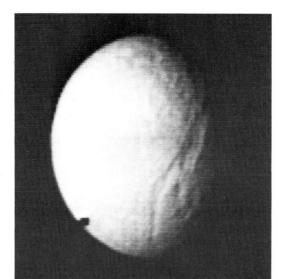

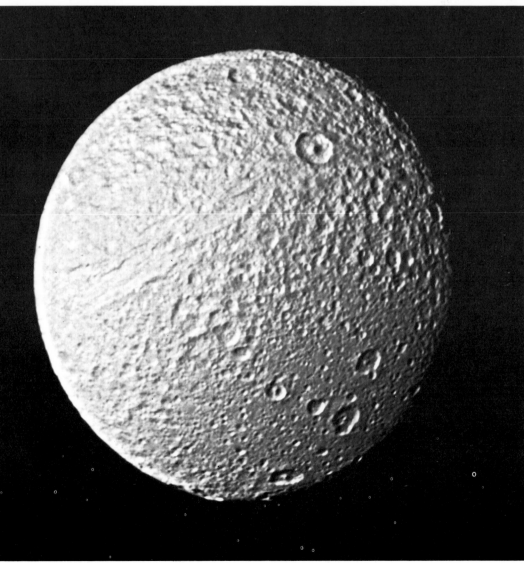

This image of Tethys shows an enormous trench some 65 km wide extending right across the moon's disc. This may be a fracture, probably related to the crater on the opposite side. Alternatively, and less likely, it may have been caused by internal mobility or expansion within the deep crust.

Map of Tethys prepared from Voyager 1 and 2
photographs. 1 degree equals 9.2 km.

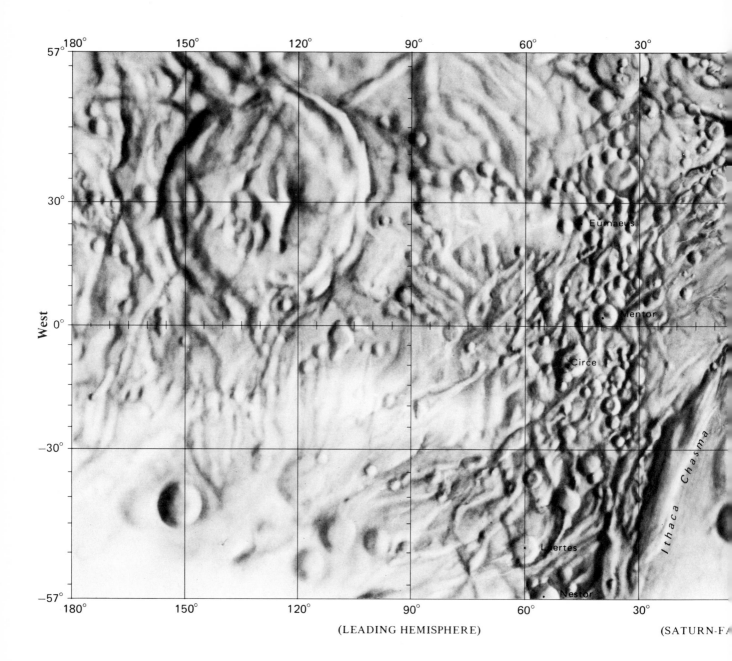

(LEADING HEMISPHERE)   (SATURN-FA

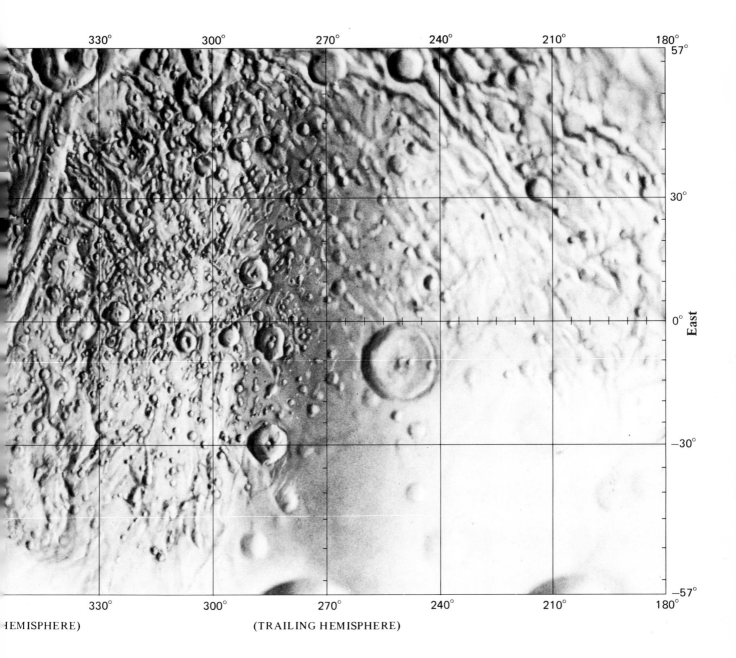

330° 300° 270° 240° 210° 180°

57°

30°

East 0°

−30°

−57°

330° 300° 270° 240° 210° 180°

HEMISPHERE) (TRAILING HEMISPHERE)

Dione is about the same size as Tethys (diameter = 1120 km) and also has a darker trailing hemisphere. Again, it is a huge sphere consisting of water ice with various other volatiles and dust mixed in, and a rocky core, perhaps 200 km in diameter. The icy surface is packed by craters caused by impacts, and by cracks caused by faults in the ice.

This distant view of Rhea has been colour-enhanced to bring out contrasts on its surface. This is the hemisphere which trails as the satellite orbits Saturn; it is darker on average than the other side. Probably, the darkening is due to a dust covering and the light streaks to fresh ice thrown over the dust as a result of meteor impacts. However, the usual 'rayed' appearance for such events is missing or subdued and implies that leaking volatiles from the interior – a sort of volcanism – may be responsible instead.

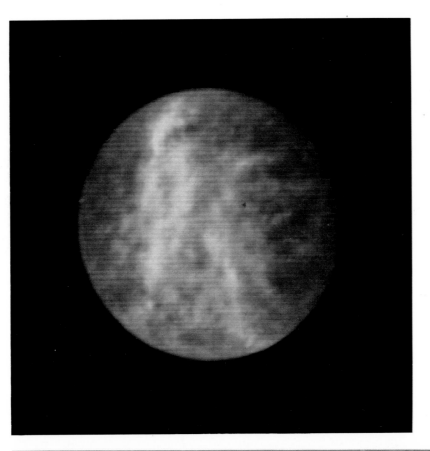

Close-up views like this of cratering on Rhea show bright walls on the edge of some of the larger craters. Near vertical surfaces may be less prone to darkening by dust deposits.

Map of Dione prepared from Voyager 1 and 2
photographs. 1 degree equals 9.8 km.

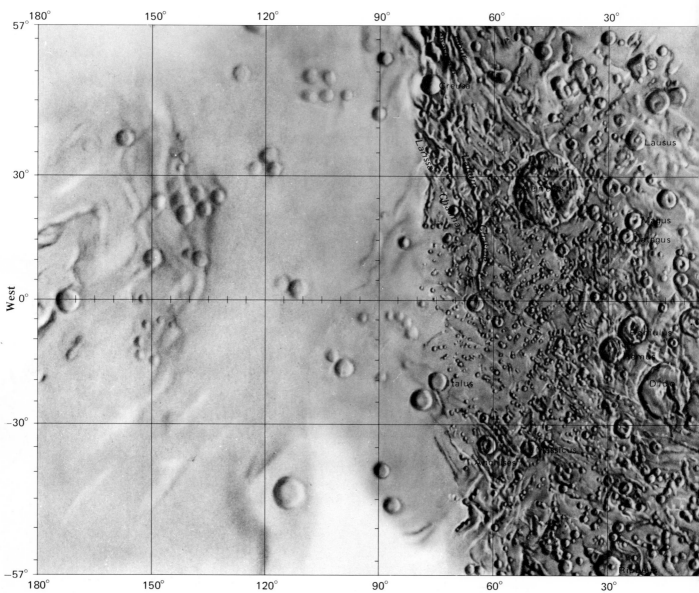

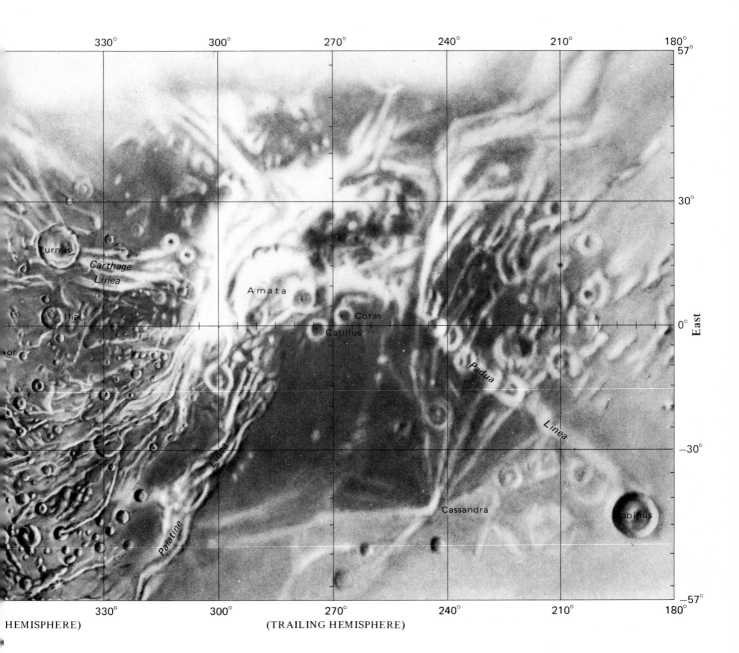

HEMISPHERE)   (TRAILING HEMISPHERE)

243

Map of Rhea prepared from Voyager 1 and 2
photographs. 1 degree equals 13.4 km.

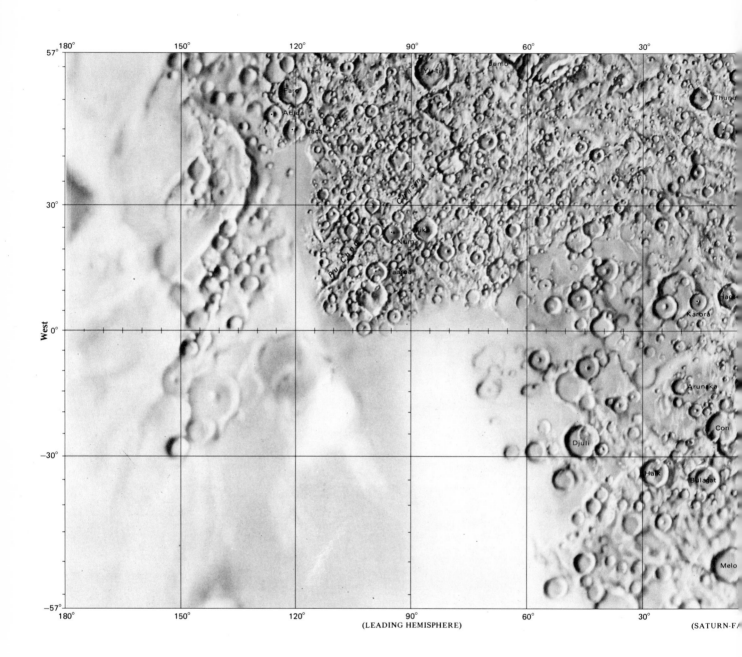

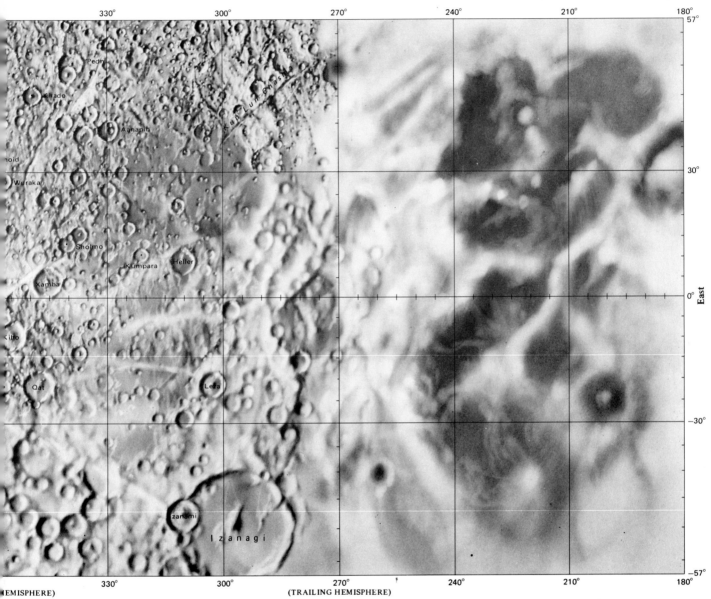

57°

330°   300°   270°   240°   210°   180°

Ped

Kiiado

Aananin

30°

oid

Wuraka

Sholmo

Kumpara   Heller

Xamba   East

0°

Kiio

Qat   Leza

-30°

Izanami

I z a n a g i

-57°
330°   300°   270°   240°   210°   180°
(EMISPHERE)   (TRAILING HEMISPHERE)

The surface of Rhea is totally saturated by craters, so that new impact events wipe out as many old craters as they create new ones on average. The region seen in this mosaic is near the north pole of the moon. Rhea is 1530 km in diameter; the largest craters are about 300 km across.

Saturn's mysterious satellite Iapetus, showing most of the bright, icy trailing hemisphere and part of the extremely dark leading hemisphere (lower right). Opinions are divided as to whether the dark material is swept up from space by Iapetus in its orbit, or whether it has oozed up from inside. In the former case, it is probably carbonaceous chondritic material from meteorites, in the latter it may be carbon from the oxidation of methane ice in the crust.

Map of Iapetus prepared from Voyager 1 and 2
photographs. 1 degree equals 12.6 km.

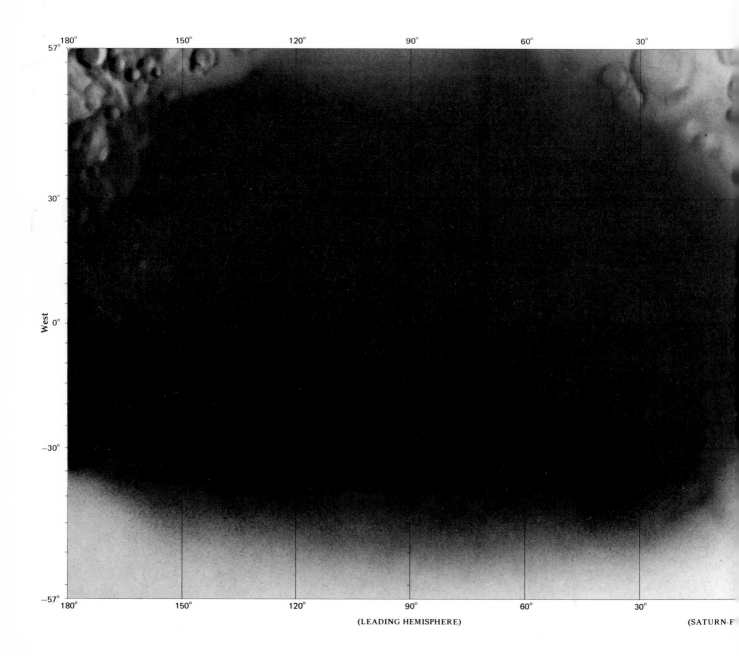

(LEADING HEMISPHERE)

(SATURN-F

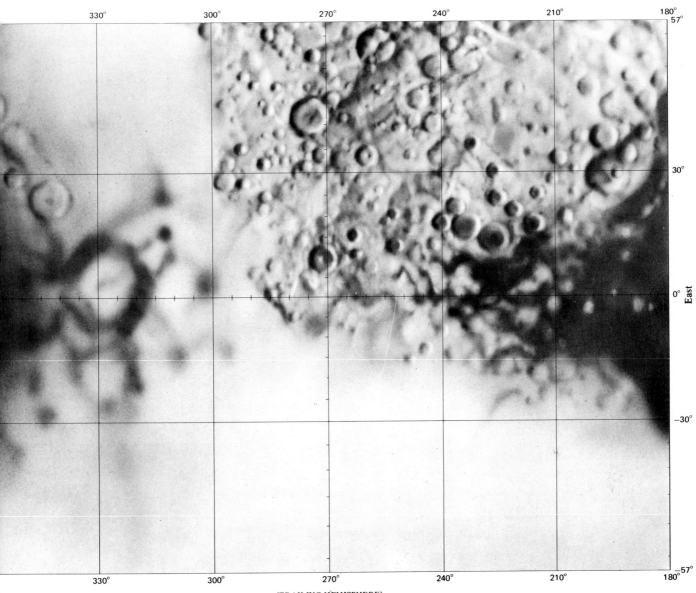

30°

0°   East

−30°

−57°

(EMISPHERE)   (TRAILING HEMISPHERE)

The Saturnian system

A Voyager 2 view of the
night side of Titan.
Only bodies with
atmospheres show this
bright ring effect,
caused by scattering of
sunlight in the clouds.

Titan, with a diameter
of 5118 km, is much
larger than Earth's
moon and only slightly
smaller than Ganymede
of Jupiter. Its most
arresting feature – the
thick, cloudy
atmosphere –
completely screens the
surface. Its yellow
colour is probably the
result of complex
hydrocarbons produced
through photolysis of
methane in the upper
atmosphere. Note that
the northern
hemisphere is darker
than the southern, and
the presence of a dark
north polar collar. This
may be evidence for
seasons on Titan.

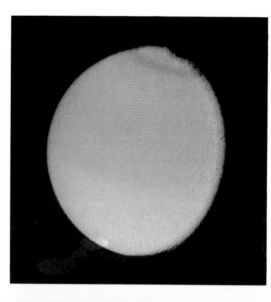

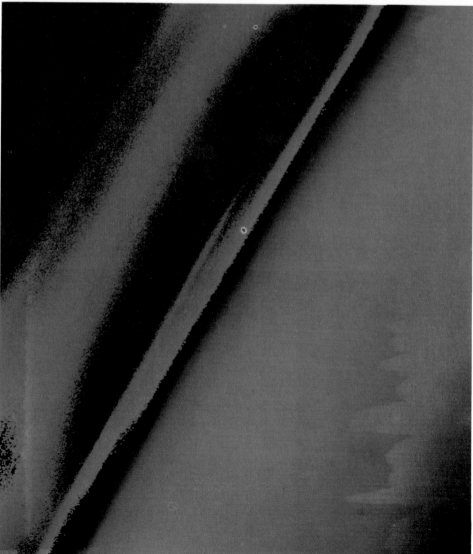

Colour-enhancement
brings out the details of
haze layers in Titan's
upper atmosphere. The
highest layer seen here
extends about 500 km
above the surface of the
satellite. The hazes are
thought to contain
organic molecules
preduced by
photochemistry from
methane. The latter is
observed
spectroscopically in
large amounts on Titan.

Saturn's moon Enceladus is only 500 km across but shows ample evidence of an interior which was active relatively recently. Some areas are cratered and others have been flooded; partial flooding of some craters near the centre of the disc is evident. Enceladus also has the brightest and whitest surface of any Saturnian satellite, further evidence of recent renewal. Its interior may be heated by tidal forces similar to those which energise the volcanoes of Io.

A series of Voyager 2 images of Hyperion, from different distances and angles, revealing its irregular shape and battered surface. Its orientation with respect to its orbit about Saturn is not gravitationally stable, perhaps reflecting a recent collision, and the moon must be either tumbling or slowly tilting back by 45° to a stable position.

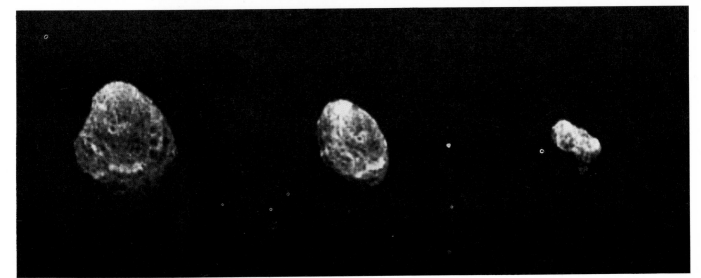

Map of Enceladus prepared from Voyager 1 and 2 photographs. 1 degree equals 4.4 km.

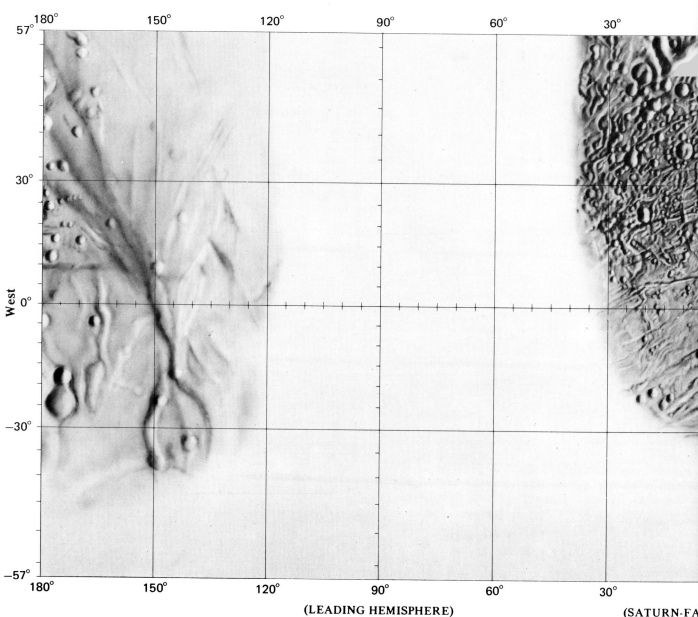

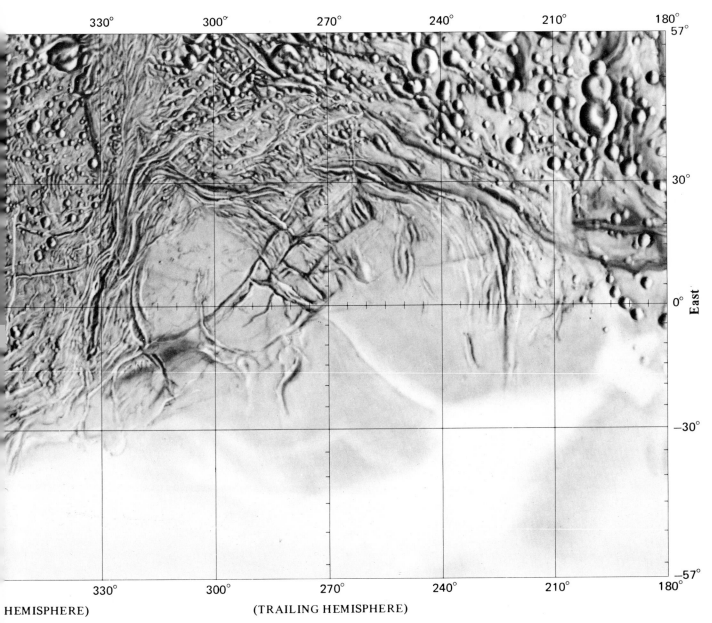

330°   300°   270°   240°   210°   180°
57°

30°

East
0°

−30°

−57°

330°   300°   270°   240°   210°   180°

HEMISPHERE)   (TRAILING HEMISPHERE)

n

No Earth-based telescope could ever obtain such a view of Saturn, since our planet is so close to the Sun we only ever see the sunlit faces of the outer planets. This picture was taken from beyond Saturn by Voyager 1 in November 1980, on its way out of the Solar System in the direction of the constellation Ophiuchus.